T0236524

SpringerBriefs in Applied Sciences and Technology

PoliMI SpringerBriefs

More information about this series at http://www.springer.com/series/11159
http://www.polimi.it

Marco Maiocchi

The Neuroscientific Basis of Successful Design

How Emotions and Perceptions Matter

POLITECNICO
DI MILANO

Springer

Marco Maiocchi
Department of Design
Polytechnic University of Milan
Milan
Italy

ISSN 2282-2577   ISSN 2282-2585   (electronic)
SpringerBriefs in Applied Sciences and Technology
ISBN 978-3-319-02800-2   ISBN 978-3-319-02801-9   (eBook)
DOI 10.1007/978-3-319-02801-9

Library of Congress Control Number: 2014956193

Springer Cham Heidelberg New York Dordrecht London

Printed on acid-free paper

Springer International Publishing AG Switzerland is part of Springer Science+Business Media
(www.springer.com)

Preface

Design, as a discipline, is rapidly evolving and expanding its field of applications and interests, out of the boundaries of traditional products and interior design domains; researchers are focusing their attention towards the development of material and nonmaterial products and systems, including communication strategies, services and industrial and social organizations. At the same time, Design theories and methodologies evolve and gain importance out of the usual domain of product design, and design cultures provide the basis to drive trans-disciplinary approaches to afford complex and troublesome problems.

In the field of Design theory, a growing attention is dedicated to recent discoveries produced by brain sciences and this new contribution of knowledge seems to influence considerably the re-thinking of project cultures and design methodologies.

There are various kinds of knowledge. Physics provides mathematical models used to predict phenomena; Economy does the same, but in the long run[1]; western Medicine builds partial models, does experiments on a disease and uses double-blind trials to verify the effectiveness of such models; eastern Medicine follows a cosmological model, based on a metaphoric symbolic knowledge; Psychology is composed of a set of different branches, each of them using a different approach, but generally using mental models,[2] and the evidence of the suitability of the model is got from the results in applications; and so on.

---

[1] Unfortunately "… In the long run we are all dead" (J.M. Keynes, A Tract on Monetary Reform, 1923, Chap. 3).

[2] Not only Psychology is using "mental models", i.e. hypothetical structures able to represent behaviours, and then to forecast; this is typical of each science; mental models are often conflictual for the same phenomenon, such as the corpuscular versus wave light models in Physics; nobody is believing that there is an observable quantity called "Superego", but the term is able to recall mental concepts, and then to explain behaviours. What we observe here is the fact that many disciplines have episthemological methods, except design, and we think that Design schools require a more rigorous episthemological approach.

In general, there are "hard" and "soft" sciences: the former, such as Physics, base their models on mathematics, while the latter use less formal models. Both are in any case characterised by models able to explain phenomena, to forecast behaviours, and, most of all, to be transferred to other individuals, to communities, to build schools.

Traditionally, Design knowledge was not characterized by specific models and the word itself is never explained in the literature of the discipline: definitions of Design are usually operational (Archer 1973; ICSID 2004; Cox 2005; Bürdek 2005; ET 2012).

The backbone of Design knowledge is better defined in terms of project methodologies and practices, and the creation of innovative products and solutions— material and immaterial—is pursued by fostering models and knowledge extracted, and in some way opportunistically adapted, by several (if not all) other disciplines such as sociology, anthropology, material sciences, mechanical engineering and so on. On the other hand, following tradition, design and more, Italian design can be also described in terms of aesthetic research, and it strives for absolute quality, pursued by masters who embodied technical skills and artistic inspiration, and who were capable to innovate shapes of objects, also bringing new meanings into daily things, such as lamps, chairs or electrical appliances.

Design deals with functions and meanings, and produces solutions valuable both in terms of satisfaction of practical needs and non-material demands; the meaning part is the most intriguing and specifically associated to Design, as discipline with no equivalent in other project disciplines. Traditionally, designers' ability to create beautiful things relies on their artistic skill: working as applied artists, designers create appealing objects capable to tickle senses and to arouse emotions; the aesthetic quality of designed artefacts is due to intuition and expertise, not to a scientific modelling of human emotional behaviours.

In the Italian tradition, masters of Design mainly counted on their own ability to achieve an optimal synthesis between aesthetic visions, technical and industrial constraints, functional needs and market requirements. Their personal culture and capability allowed them to deal with the complexity involved in the design process of objects and solutions, and the creative process could be implicit and mainly individual. Their professional capability, and the fertile cooperation they established with captains of industry after the Second World War sustained a new flourishing of Italian industry, influenced consumer tastes and behaviours, and finally inspired generations of designers all over the world. The masters of Italian design were not scientists: some were architects and some had art or other similar background education; they *signed* their pieces of work, and felt responsible for the final quality as they expressed themselves through it as artists do. Through the shape of material objects of daily use, such as tables, chairs or domestic appliances, designers broadcasted conceptual memes and metaphors, giving a contribution to social innovation, shaping tastes, setting trends. These designers were Masters, able to nurture disciples, more than to create schools.

Progressively, Design as a profession developed into a discipline with dedicated university courses, Ph.D. research activities and scientific publishing; progressively,

design as a practice moved from authorship to teamwork and co-design methodologies became more robust and rigorous, also including contributions coming from
a growing number of disciplines: ergonomics, sociology, anthropology, psychology, brain sciences. Design is not yet a scientific discipline, and in large part it is
still based on experiential knowledge. But designers now deal with multidisci-
plinary project teams and, especially for immaterial solutions (such as in the case of
service and of experience design), the final shape—and therefore, the final meanings carried by the solution—depend on a number of factors, difficult to keep under
control. So, in order to produce meaningful innovative design, designers must
upgrade the way they work and their knowledge.

A research on any kind of subject should be aimed at the construction of models
allowing to:

- understand phenomena;
- forecast phenomena;
- invent phenomena.

The phenomena of interest for designers are those connected to human activities
and human well-being. Human activities and well-being are considered in Design
from a social and collective point of view, but also in terms of individual perception, and so in terms of subjective, personal, specific and not necessarily
shareable perspectives. From this point of view, Design research must provide
models able to explain the phenomena related to human experience in its generality,
but also to cope with individual variability. Anyway, Design research should
produce knowledge useful for design practice and should always address toward a
search for quality.

Recent scientific discoveries in the field of brain and perception sciences (and
their popularization through some books unveiling their importance, especially in
marketing) are arousing growing attention in the research of several disciplines such
as Economics, Marketing, Communication.

The main reason for this popular interest is quite obvious: neurosciences, brain
mapping activities, modelling of decision processes provide highlighting on the
most hidden and unconscious mental processes; automatic decision processes
become less mysterious; individual and collective perception phenomena find new
convincing explanations.

However, scientific models do not produce in themselves good design as the
knowledge of Gestalt laws is not enough to guarantee good graphic design. So the
question to discuss now is how Design can include brain sciences into design
practice, without losing its soul.

Thanks to the contribution of new knowledge provided by brain scientists,
Design faces great opportunities to evolve. This evolution is mandatory as design
practice faces great changes from the technological, social, economical and
industrial points of view.

Brain sciences explaining cognitive and emotional processes can be introduced
into the project culture in several different ways and for a number of reasons. In

some recent works [f.i. Norman (2004) and Van Gorp (2012)], Design methodologies are treated as kitchen recipes, including tools, techniques and procedures including prescribed activities; thanks to models provided by neurosciences, the complexity related to human emotions and cognitive processes are reduced to handy constraints to be taken into account through purposely ideated activities, but this approach, implicitly inspired by a reductionist vision of the world, is not the most promising; on the contrary, it seems to reflect an undervaluation or, worst, a miscomprehension of the original and irreplaceable role of Design as a specific form of anthropology and a specific facet of epistemological praxis.

The present book resumes the results obtained during several years of research activity carried on by the author; the research was based on theoretical work and experimental design and prototyping of innovative concepts of product and services.

In detail, our fields of interest include:

- brain sciences as a basic knowledge, useful in modelling cognitive and emotional mental processes activated by products, services and communication created by design;
- the role of brain sciences in the upgrade of design methodologies, in order to support the design of complex artefacts, and mainly of non-tangible solutions, as digital services and systems;
- the importance of knowledge provided by brain sciences with respect to creativity, awareness and responsibility in the design of innovation.

The present book intends to give a contribution to Design theory, exploring the importance of new scientific discoveries to support innovative design practices, but also emphasising the need for continuity with respect to authorial work based on individual intuition.

While we were writing this book, almost every day we found news in the newspaper indicating the growing importance of brain sciences in every field of project: automatic interpretation of facial expressions are used to measure and analyse the involvement and emotional reaction in front of movie trailers; sentiment analysis on big data flows provides accurate forecast about political elections; computer engineers design safety systems based on emotion analysis systems; window dressers of shopping malls orient their activity on the basis of affective computing software application, made available for reasonable prices.

In the light of these changes, it is only natural to ask ourselves about the future of Design as a discipline able to provide messages, contents, emotions through the shape of material and non-material products and services and, mainly, through the creation of experience. Will engineers together with scientists definitely substitute Design? Will brain sciences and big data eliminate the need for poetical inspiration? Will functional Magnetic Resonance remove the need for creativity? How will Design education change in the light of new knowledge?

This book is a contribution to the research for answers to these questions, but also an attempt to reinforce the cultural continuity between old and new design approaches.

Without laying any claim to providing a scientific background to this discipline, nor to replacing the needs for creativity, we believe that this book could give some contribution to the enlargement of the horizon of designer practice, as well as to take a little bit farther the boundaries of the discipline knowledge.

While it is essential to update the discipline in the light of new scientific knowledge, it is also essential to think back on its specific finalities and methodologies, and also to develop a debate on the new ethical issues and professional responsibilities provided by the inclusion of brain sciences discoveries into design practice, also learning from Italian design tradition, where designers signed their pieces of work, and fully assumed the responsibility of practical and cultural consequences of their actions.

The Book

Design is not a scientific discipline. Design can be defined in a number of different ways and the goal of reducing the amplitude of the existing definition into a unique statement is a useless task. As a start, we will assume that, following the tradition of Italian Design history, Design is applied art, or a discipline aiming at making more robust the activity of designing solutions to problems. The invention of new solutions to practical and non-material problems and needs is quite a natural activity for human beings, but Design can provide tools, methodologies and a cultural background supporting projects of more qualified objects and services. The quality refers to the functional and formal (aesthetic) attributes.

Design deals with human needs, wishes, attitudes, tastes, potentialities, cognitions, culture, so exploiting every form of anthropology. Furthermore, Design deals with technology, and the role of Design is the invention of technology applications, aiming at the fulfillment of human satisfaction in the amplest sense.

Design is changing since human societies and values, and cultures, and organizations are changing. The main changes in Design can be listed in terms of:

- from material products to non-tangible goods, from objects to services;
- from static things to interactive solutions;
- from artistic authorship to teamwork and co-design;
- from the aesthetics of material shapes to user experience;
- from Design as style counselor orienting tastes, to Design for Social Innovation.

As Design changes, we need to update Design theory and methodologies. Our contributions to the upgrade of Design theory exploits knowledge provided by brain sciences and it is based on the proper use of this knowledge for a number of purposes.

Our first aim is a sounder explanation of the mechanisms of attraction exerted by objects coming out of traditional design and being qualified in terms of shape and appearance. Our thesis is that the scientific knowledge of emotional thinking, and notably Panksepp results, modelling emotions as primary activities mainly of the limbic system, can be suitably employed to analyse the correspondence between material (mainly visual) attributes of Design objects in terms of their capability to arouse primitive emotions in users. This kind of analysis produces new awareness with respect to the so-called formal qualities of Design objects, induces new readings of the social relevance of design as an activity aimed at vehicle contents, messages and values through implicit communication related to formal attributes. Furthermore, a greater awareness of the communication consequences of emotional outcomes of design choices makes design activity more powerful and therefore more responsible.

Our second aim is to demonstrate how brain science results can contribute to update design methodologies, providing conceptual tools useful in all project phases: from preliminary ethnographic analysis to concept generation, to prototype and evaluation tests to team building and management in multicultural and multi-disciplinary project groups. To these purposes, Design can draw a number of useful models of emotional thinking, referring to different authors making research in the field of brain sciences: besides the above-mentioned Jaak Panksepp, in the book we extract useful knowledge from the works of Vilayanur Ramachandran, Marianella Sclavi and others, whose writings have been precious to us during the last years. We mainly focused on an Interaction Design, which is the application domain we operate in most, but results could be extended to all other fields of project activity.

Finally, the third part of this book presents some design experiences conducted following the design approaches illustrated in Part 2 and employing brain science knowledge as conceptual tools useful in the project of practical solutions. These case studies illustrate how traditional design can be empowered by new knowledge, enforcing creativity and increasing the importance of Design-specific contribution in the field of project cultures.

Conflict of interest The author reports no conflict of interest.

Acknowledgments

This book is the result of a long research carried on with many contributions. First, teaching and proposing challenging exercises to my students provided a lot of material and suggestions able to drive explorations and hypotheses; many examples and experiments are due to graduation theses: thanks specifically to Humberto Sica, Monica Diani and many others; the discussions with my Politecnico researchers were hugely fruitful: thanks in particular to Francesco Galli, Marko Radeta and Zhabiz Shafieyoun. Many friends and colleagues supported, through suggestions and criticisms, the research done: many thanks to Giovanni Cutolo, Gabriele Foglia and Francesco Trabucco. Thanks to Cynthia Ortega for many illuminating discussions and works together. Thanks to Dijon de Moraes for the discussions and the international opportunities provided. Thanks to Jaak Panksepp for the clean though short discussions, supporting the research. Thanks to Decio Carugati for the intelligent discussions during many pleasant dinners. Thanks to Anna Maria Fattore for the diligent reading and the deep and wide discussions.

But, for sure, this book is the result of a pluriannual tight cooperation with Margherita Pillan: she provided me with fruitful support, intelligent criticism, deep design culture, suggestions, always accepting any foolish proposal, and turning it into good opportunities for challenging researches and results. The concepts and the theoretical aspects expressed in this book are totally the result of this cooperation.

More recently, thanks to Luca Guerrini, Alessandro Biamonti, Maurizio Mauri, Alessandra Orlandi, Gabriella Pravettoni, who determined, possibly unconsciously, the completion of this book.

Finally, thanks to my affectionate wife Youngju and to my daughters Delia and Beatrice, who gave me, besides really useful suggestions and hints, also the serenity to carry on the research summarized in this few pages.

These acknowledgements are once more the demonstration that any very small step in knowledge is allowed by the fact that we are climbing on the shoulders of giants. Not necessarily giants in the same discipline.

Contents

Chapter 1
Emotions and Design Methodologies

Abstract The chapter discusses about the relevance of the emotions in designing artefacts, and points out the chance offered by the neurosciences today, in order to give sound basis for studies and methods for Emotional Design. The goals as well as the limits of the book are presented.

Keywords Emotional design · Brain sciences

There are at least two different approaches to design: one is technology driven and the other is mainly focused on non-material needs and toward aesthetic qualification. Since industrial revolution, Industrial Design has been coping with both, trying to satisfy functional and emotional needs by designing useful industrial objects also capable to carry meanings and metaphors through shape, colours, materials. After Second World War, Walter Gropius and the Bauhaus movement were quite conscious of the importance of good design of daily objects: through the democratic qualification of popular products they saw the opportunity to dignify people and life.

Italian Design with its famous masters such as Vico Magistretti, Achille Castiglioni, Enzo Mari, had the capability to envision new aesthetic paradigms, so making modernity attractive and glamorous. The economic relevance of this contribution in the industry system development is evident and notorious, but still the potentialities of emotional Design are underrated.

We face a systematic undervaluation of the importance of non-material needs with respect to the practical ones and tend to consider emotions as the weak part of our existence, while they are fundamental components of our intelligence; emotional and intuition-based process are crucial and much more relevant then what we usually tend to believe.

According to many authors (Darwin 1872; Plutchik 2002; Ekman 2003), emotions are the status related to processes associated to the evaluation of given situations, or, in other words, reactions to external events that our brain consider as relevant for our wellbeing. When we perceive that something significant for our life is happening, our body reacts as a consequence of the evaluation performed by the brain. The evaluation depends on our individual and personal history as well as on the whole human-kind experience resulting from evolution and natural selection. As cognitive psychology and neurosciences develop and build deeper and more complex models of human brain and intelligence, the importance attributed to the non-conscious thought grows and its importance in all aspects of our life is recognized. Visceral and behavioural mental processes, as Donald Norman call them (Norman 2004), drive our car while we are engaged in a conversation using our reflective mind in a conversation with the passenger nearby us; thanks to them, we can quickly react if some obstacle or problem interfere with our trip, recalling our full attention to the new emergency. They make us play tennis or skiing or do any kind of complex tasks, allowing much faster reactions then those connected to our fully conscious metal processes.

Non reflective mental process are quite capable to judge situations, people and contexts; they manage our senses to exchange information, evaluating minor signals and multi-sensory background effects (Rosenbaum 2010), comparing them we memories of past experiences, and effectively sorting decisions through non linguistic evaluation systems.

When our perception system (senses + mental processing + evaluation) detects significant events in the environment around us, our body reacts to prepare for action by increasing heart beat rate and blood influx into arts and brain, but also producing expressions on our faces. Rage, sadness, disgust, surprise, pleasure, fear, ... all them manifest themselves on our expressions and are easily recognized by others thanks to mirror neurons providing the biological base for empathy (Rizzolatti 2008). The recognizable evidences of our emotions support communication, allow knowledge sharing and pose the base for social relationships.

Emotions are something we can only partly control with our reflective mind, and all of us have a contradictory relationship with our ungovernable and irrepressible emotional reactions since quite often we fill that they interfere with our conscious willing. We dislike dissonant mental processes, but our mind constantly works through complex dynamics, and through internal conflicts between competing cognitions, goals and priorities (Festinger 1957; Cooper 2007).

Design has always coped with the complexity of human mind and, playing with sensorial experiences (with colours, material, shapes) addresses to the inner recesses of human minds lighting desire and pleasure.

The ability to arouse emotions through artefacts is the main recognized skill for artists and designers; traditionally, this ability is considered as a soft and fluid competence, it is mainly based on personal sensitivity and it is transferred from master to apprentice together with technical knowledge in informal methodology and empirical practices.

On the other hand, marketing as a discipline has been exploiting the scientific knowledge on human mind since the beginning of the twentieth century, when the first psychology theories began to be employed to better understand purchase behaviours. After Second World War, thanks to the intense developments of psychology studies, motivation analysis provided the guidelines to develop more attractive products and communication, so offering a significant contribution to industry development (Packard 1957). More recently, marketing experts do not hesitate to employ brain mapping based on FMR-Functional Magnetic Resonance, to fully understand deep emotional processes related to product consumption so to better understand contradictory and un-admitted feelings, desires, passions (Lindstrom 2010) and optimize products and increase consumptions.

The scientific understanding of human behaviours and mental processes have been employed to innovate products and services, but the knowledge of human brain processes is usually associated to negative purposes of manipulation and hidden persuasion.

Anyway, this knowledge it is still underexploited by the Design community that tend to prefer traditional approaches mainly based on social studies and design practice.

Basing on our Design works and researchers, we became convinced that a deeper knowledge of brain sciences and mind processes can give a contribution to Design and to designers in terms of:

- increase of sensitiveness toward sensorial and cognitive experiences;
- strategic meta-design activities, suitable to drive project early stages and new concept generation;
- increase of creative ability.

In the following we will discuss these statements basing on examples, case studies and past research experiences.

Chapter 2
Design as Evolutionary Discipline

Abstract The chapter discusses the definition of what Design is; an historical perspective is provided, suggesting which are the relevant aspects of Design vs Industrial Design; a formal model of Design is then provided.

Keywords Design definition · Design evolution · Design model

2.1 What Design Is

Many definitions of Design have been provided in literature, but most of them are related either to the characteristics of the artefacts or of the involved processes.
Cox, Chairman of the British Design Council, defines Design as (Cox 2005):

> Design is what links creativity and innovation. It shapes ideas to become practical and attractive propositions for users or customers. Design may be described as creativity deployed to a specific end.

but this definition is useless both for driving the designer's activity and the artefact properties. Other definitions are present in literature, such as the one provided in 1979 by the *Internationales Design Zentrum Berlin*:

- Good design may not be a mere envelopment technique. It must express the individuality of the product in question through appropriate fashioning.
- It must make the function of the product, its application, plainly visible so that it can be understood clearly by the user.
- Good design must allow the latest state of technical development to become transparent.
- Design must not be restricted just to the product itself; it must also take into consideration issues of ecology, energy conservation, recyclability, durability, and ergonomics.

M. Maiocchi, *The Neuroscientific Basis of Successful Design*,
PoliMI SpringerBriefs, DOI 10.1007/978-3-319-02801-9_2

- Good design must take the relationship between humans and objects as the point of departure for the shapes it uses, especially taking into account aspects of occupational medicine and perception.

This statement too describes what Design should cope with, but does not define what Design is. Further definitions do not provide any answer to our question: what is Design?

The question is not simply the matter of a definition: only by understanding and sharing a possible answer we can define the foundations of a discipline, and then what should and should not be part of the educational process for Designers.

In our mind, Design is a specific activity, able to give specific properties to the projected artefacts, activating specific perceptions into the users: we think we must indicate what the above specific activities, specific properties and specific perceptions are; from them we can identify goals and conceptual tools for the designers and the disciplinary contents to be learned in schools and universities.

In order to better understand the role and the meaning of this discipline, we try to examine it from the point of view of economy history.

2.2 Craftsmanship, Industrial Design, Design

At the beginning there was craftsmanship; the craftsman was able to build fine and precious object, for the day by day use, not necessarily as exclusive luxury goods. Beside the production of jewels, the same working process was in principle applied to swords, knives, ploughs, carts, pots, and so on.

Craftsmanship was characterised by the craftsman capabilities: experience, product competence, apprenticeship, non-formalised excellence, one shot products, personalization, and the products had high costs.

For example, when a Japanese craftsman was producing a sword, the result in terms of quality was related to his experience, sensitivity, aesthetic feeling, etc., and he was able to provide in the long run and at very high costs an excellence product. Each sword was a unique object, properly built according to the personal needs of the customer (size, weight, shape, aspect, symbols, decorations, and so on); the building process was driven by taking care of signals, difficult to explain and formalise (the sound while beating, the smell while tempering, and so on); the impossibility of formalising all those elements prevented from schools set up, and the only way to pass on hand down the know how was just apprenticeship.

With the growth of middle class, the requests for goods required a larger production capability and a reduction of costs: industrialisation can be seen as a consequence of those facts. Capitals, previously used just for financing wars or for buying lands, went to industrial plants creation. For this purpose, the possibility of using non specialised man power was very important, that is, transferring the capability of the craftsmen into well defined production phases, able to guarantee the final result, no matter who was executing them. The need of a standard quality and of a huge product numbers resulted into serial production, with attention to processes and to their definition, and in particular to the initial design phases: Industrial Design was born, with attention to the early phase,[1] to formalised and measured processes,[2] to standardisation, to cost reduction and to controlled quality.[3]

The serial production method spread everywhere, and competition increased: today, also complex products as cars, are, for each market segment, aligned from the point of view of costs, performances, consumptions, maintainability, life duration, and so on.

The only way a maker had to compete has been to add some new "meanings" to the products, making the difference, and adding some reasons for convincing customers to buy one product instead of another: when a potential user enters a shop to buy specific goods, for example a microwave oven, he/she will find a lot of similar products, with the same technical specifications, with aligned prices, and the choice cannot be made simply rationally; the winner will be the product meeting unnecessary expectations (the colour, the shape, the sound of opening/closing the door, and so on).

For the above reason, besides the proper functions and an acceptable cost, products started to present shapes and styles able to communicate to users emotions; it is what we now call *Design*: industrial production assuring costs and quality is considered unavoidable and implicitly understood, while a product must have the capability of communicating some kind of emotions.[4] In some way we can consider the present status of mass products industry as in a post-industrial phase, where capitals are, in some way, a commodity.

So, when we think of a Design artefact, we should mean an element in which at least three different "components" are present: *function*, *shape* and *meaning* (Maiocchi et al. 2009):

- *Function* is the goal for which the artefact is produced: a vacuum cleaner must suck, a squashing machine must squash, a car must run, and so on; the basic

---

[1] A project error occurring in a phase has recovery costs growing more and more as the error is discovered too late; a design error that can be recovered with cost 1 if recognized immediately, could cost 10 if discovered during the prototyping phase, 100 if discovered during production, 1,000 after distribution. The early phases have the goal to prevent any kind of problem before we start to develop.

[2] Formalisation guarantees the constancy of the process and the absence of ambiguity in communicating how-to-do; measures allow to define quantitative goals in production improvements (more pieces per time, stricter tolerance parameters, and so on).

[3] What is usually called quality control is intended as an activity related to some pass-no pass check for proceeding to a further working phase, but it is the technical and cultural basis for the continuous process improvement, often referred as Total Quality.

[4] By the way, everybody is able today, also when non-expert, to locate on a time scale the production of a disc player or a radio, just by looking at it; exterior perceptual properties reveal the social models, the aesthetic values of a specific period, and so on; besides the functions, an object hat to identify itself in a precise social context, and this goal was left to what we call today Design.

function should be efficient and effective, and among functions we consider also what is relevant to the use (e.g., complementary functions, ergonomics, affordance, maintainability, etc.).

- *Shape* is the geometrical appearance: from a structural point of view, it should make an artefact recognisable as belonging to a specific archetype (e.g., a vacuum cleaner is basically made of a body, a tube and some brushes), but geometric details can suggest different interpretations, carrying in itself some meanings (round shapes or sharp edges can attract or repel customers); in a broader sense, we consider "shape" any kind of appearance for any kind of sensorial channel, such as hearing, touching, smelling, tasting; the same vacuum cleaner in white metal or coloured in dark blue, bright pink or purple can change completely its soul, and be accepted/preferred by different customers.
- *Meaning* is the communicated interpretation, often unconscious and often completely unrelated to the functions, able to carry with it emotions; a proper set of shapes can make a luxury item of an artefact, a *retro* or a *hi-tech* object, an exotic element, elegant, for men, for specific stereotypical persons, and so on.

Let us take a look at the following picture, showing many kinds of vacuum cleaners (Fig. 2.1):

The first is one of the archetypical products of the Fifties, and its structure is usually recognisable in the more recent ones, also when hybridised with a broom (as the last one); some of them change the archetype (second and third in the second

Fig. 2.1 Different vacuum cleaners

row), and disguised as intelligent hi-tech object (possibly for upper-class professionals living in a loft); some of them use irony, winking to play (the last in the second row and the snail in the third row); the *Bidone Aspiratutto* (last in the first row), talks to us about power and war, transforming the housewife into a soldier against dirt; the Dyson model (first in the last row) suggests a sophisticated aero spatial technology, transforming the user into a *Ghostbuster* of the powder. For sure the buyer's choice is influenced by the meaning communicated through shapes.

In fact, the communication with humans does not take place as most of the semiotic manuals present: it is not the flow of some code information from a source to a receiver through a channel (model apt to describe a technical transmission as by radio waves); communication should be instead considered as the action to put a new information into a complex context of knowledge interconnected through many relationships: the context contains a lot of elements, parts shared by both source and receiver, and part known only to one of the parts; moreover, those items are connected by relationships, again shared or not; by adding a new item, the source compels the receiver to re-organize his network, possibly creating new meanings. This is what we think happen when we provide a good design artefact: a new information appears within a (metaphoric) network of connected concepts, and new meanings are automatically connected to the new item, providing it with new interpretations, and possibly with new emotions (Fig. 2.2).

Just in this way, the use of military green in the *Bidone Aspiratutto*, with military stencil in white, the rough shape and the essentiality of the elements are pieces of communication; the receiver connects those elements to the vacuum cleaner function, and the new relationship network leads metaphorically to the interpretation of the power of the product, to the role of the army, and transfers such a power to him/herself, as a new *Rambo* provided with a strong weapon. In the same way, the white metallic varnishing of the plastic body of Dyson's model, moulded with sticking out parts recalling science fiction, together with the rolling up of its pipes, suggests a strongly sophisticated and futuristic object, whose effectiveness is certainly linked to the high level technologies employed in the project and in the production of it.

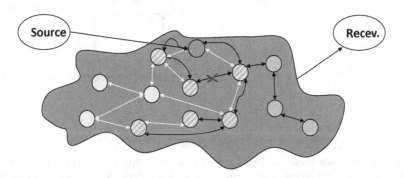

Fig. 2.2 A model of communication

2.3 A Model for Design

We consider the following scheme[5] a suitable model for Design activity (Fig. 2.3):
In the model, besides obvious technical (including economic) constraints:

- *Signals*: any kind of information, by any kind of channel: shapes, colours, sounds, flavours, words, smiles, but also proportions, big eyes, curves typical of "puppyness", pentatonic musical scale, "bi-stable" ambiguous signals, and so on.
- *Meanings/emotions*: the results of the processing of the gathered information; for example, some of the signals can suggest maternal feelings, exotic environments, or simply pleasure.
- *Simple perception*: processes activating the more ancient parts of our brain, where primary emotions arise (seeking, sex, fear, social interaction and approval, etc.).
- *Complex perception*: processes activating the neo-cortex (logical and cognitive functions), interacting with the lower levels, then providing emotions.
- *Cultural constraints*: besides the basic primary feelings, common to every man (and often to apes, mammals, chickens, etc.), there are inherited cultural aspects that allow or prevent the acceptability of some values (and then the associated metaphors); in other words, an artefact expressing some meanings, suitable for a Japanese, could be unacceptable for a Dutch.

In the following chapters we will examine more deeply some of the above concepts, and the related implications to our Design model.

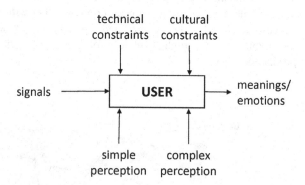

Fig. 2.3 A model for design activity

---

[5] The used formalisms known as IDEF0, presenting a process (the box), with horizontal arrows representing input and output, lower vertical arrows representing mechanisms activating the process, and upper vertical arrows representing constraints.

Chapter 3
Emotions and Design

Abstract In this chapter, the structure of the human emotional system is presented, and the seven main emotions, according to Panksepp's work, are discussed; the triune brain theory is introduced, and the different emotions arousal is discussed; many examples are provided, applied to design, communication and interaction design.

Keyword Emotions · Triune brain · Reptilian brain · Old mammalian brain · Neo cortex · Seeking · Rage · Fear · Lust · Care · Panic · Grief · Care

3.1 Emotional System

According to most of neuroscientists, the human brain can be schematized into three layers, referred to the seniority in the species evolution. The inner part (*reptilian*) is the site in which primary emotions arise, mainly related to survival (seeking, fear, anger, etc.), the middle part, developed in mammals, is related to some typical maternal and social emotions, and the upper part (*neo-cortex*) is more related to rational and logic processes (Fig. 3.1).

Among the various available models describing the physiological mechanisms of emotions, we selected the one developed by Panksepp (2012).[1]

The reasons for this choice are its simplicity and its perspective:

---

[1] According to Panksepp, each emotion rises in a specific part of the brain, and is connected to specific neurotransmitter circuits; in this way, it is possible to define scientifically what an emotion is, and to have a well founded taxonomy of emotions.

© The Author(s) 2015
M. Maiocchi, *The Neuroscientific Basis of Successful Design*,
PoliMI SpringerBriefs, DOI 10.1007/978-3-319-02801-9_3

- Panksepp considers emotions as one of the mechanisms invented by evolution to increase the survival probability of the species; according to it, animals feel emotions as we humans do; Darwin was the first considering this emotions role (Darwin 1872), and most of the researchers are going to accept this view.[2]
- Panksepp considers an emotion as the activation of a specific part of the brain through a specific neurotransmitter class, and the number of basic emotions emerging from this approach is of only seven.

Nevertheless, the works published by Panksepp don't examine in deep the peripheral nervous system (sympathetic and parasympathetic systems, vagus nerve, etc.); those aspects are handled by other authors, such as Sapolsky (2004) and Porges (2011).

Keeping into account both the aspects, it seems obvious to build a model in which:

- there are seven basic emotions, all of them rising in the "inner part"[3] of the brain: Seeking, Fear, Rage, Lust, Care, Panic/Grief, Play;
- those emotions are able to influence the physical state of the entire body (blood pressure, heart beat, sweat, etc.), that we join to emotions feeling;
- the central and the peripheral systems are possibly non connected in a hierarchic way, and emotions can be raised starting from anyone of them.

The various parts of the brain are interconnected and are able to influence each other. For instance, if our neo-cortex is listening to a tale in which some events recall the family and babies, the reorganised metaphoric structure can influence the

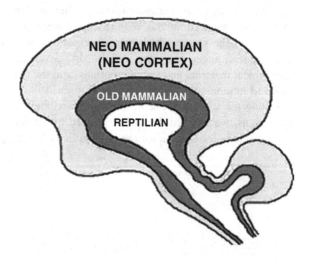

Fig. 3.1 A scheme of the human brain structure

---

[2] Damasio, LeDoux and many others (private communication with J. Panksepp).

[3] According to the scheme of a three-layered brain. reptilian, old mammalian (or limbic) and neo-cortex, the emotions are related only to the first two levels.

intermediate part of the brain, with the production of neurotransmitter related to maternal emotions, and, depending on the events of the tale, also the reptilian part, inducing fear.

Following Panksepp, there are seven basic emotional systems:

- *Seeking*: makes creatures interested in exploring, and in getting excited when they get what they might desire.
- *Rage*: aroused by frustration, tends to freedom of action.
- *Fear*: leads creatures to run away, or, when weakly stimulated, to freeze.
- *Panic/Grief*: governs social attachment emotion, specifically for the absence of maternal care when babies.
- *Lust*: involves sex and sexual desire.
- *Care*: maternal love and caring.
- *Play*: pushes young creatures to facilitate learning.

As Panksepp suggests, *seeking* can be considered the mother of the other emotions, in the sense that, while it constitutes a primary impulse pushing us to look, search, investigate, in order to give a meaning to what is surrounding us, not only it is the basis to find what can satisfy physical needs (hunger, thirst, sex, …), but it is activated also by more abstract concepts (shelter, danger absence, …), or even by purely abstract (ideas, models, forecasts, …) (Fig. 3.2).[4]

Fig. 3.2 The emotions

---

[4] Seeking can be recognized, for example, also in the pleasure while solving clues, or crosswords, or other kinds of enigma. The pleasure provided through seeking is related to dopaminergic brain circuits, mainly located in the so called *Nucleus Accumbens*, the same area excited by many drugs, such as opium.

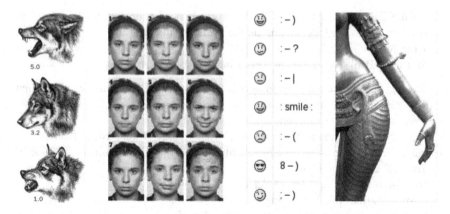

Fig. 3.3 Seeking is what allow us to interpret the attitude of a wild animal, just by details (such as the ears position or the teeth), or the psychological status of a person; it activates meanings from symbols, and connects different details into conceptual networks (such as female shape, plus refinement, plus sophisticated decorations and jewels as situation in which *lust* and *play* can be simultaneously activated)

Seeking is what allow us to recognize very quickly (without any conscious process) the context, reacting immediately to a possible danger, or to positive signals, involving in that other emotional circuits, such as *fear*, *rage*, *lust*, and so on (Fig. 3.3).

The Seeking allows also to decode emotions perceived by other individuals, f.i. perceiving Rage of Fear in others. This ability is related to the fact that the emotions influence the body through the peripheral nervous system, and then its expressions, postures, gestures, muscles behaviours.[5] The chain is then (through an example):

- subject X, male, activates Seeking, i.e. the central nervous system prepares the peripheral system to useful behaviours (muscles ready to attack, or to escape, attention by eyes or ears, etc.);
- subject X sees another individual Y, male of its same species; in order to protect his territory, Rage raises, and this influences the peripheral system, producing adrenaline, and providing aggressive postures and expressions;
- Y sees X, its Seeking decodes postures and expressions of X raising either Rage or Fear: the prevailing emotion will determine the following postures, expressions and behaviours (fight or flight).

Those mechanisms get embodied in the DNA of the species, as the capability to recognise patterns, that become stereotypes; possibly, those stereotypes are recognised as static frames, and not as stories (i.e. erect ears and shown teeth are a picture, and not a movie).

---

[5] See also the works by Rizzolatti about the "mirror neurons".

Fig. 3.4 Seeking can connect the same signals in different networks, activating alternatively *fear* or *care*

All those aspects, relevant both for the inter-species and intra-species relationships are the basis, for the latter, of social behaviours.[6]

The Seeking construction of the conceptual networks is able to dramatically change the meaning of similar signals, as shown in Fig. 3.4.

It is evident as Seeking provides taxonomies and stereotypes, possibly related to embedded brain structures; beside such a kind of primary classifications, others take place, on a metaphoric basis[7] (Lakoff and Johnson 1980) ; the relative paths from perception to emotions for the two pictures of Fig. 3.4 are quite different: while the former can skip conscious processes and stimulate directly the lower brain levels, the latter activate stereotypical metaphors, able to drive interpretations able to stimulate emotions, as shown in Fig. 3.5.

Panksepp does not introduce a clear hierarchy among the various emotions (but Seeking), but it seems natural to introduce the following:

1. Seeking: is the basis for the interaction with the world, allowing to rise other emotions.
2. Lust, Rage and Fear are common to any species of Chordata (i.e. where there is something at least similar to a brain); they are the basis for survival through reproduction, nutrition, aggression, defence, escape.

---

[6] In a broad sense, may be that the capability of a car seller to understand the psychology of a buyer has the same roots.

[7] Following Lakoff's work, as *metaphor* we intend the operation of a (partial) overlapping of two (formalisable) semantic structures, so that many of the relationships among the items of one field are the same of the ones taking place among different objects in another field. To say "I cannot digest your words" is based on the overlapping structures *food–to eat–stomach–digestion*–etc. and *words–to hear–mind–acceptance*–etc.

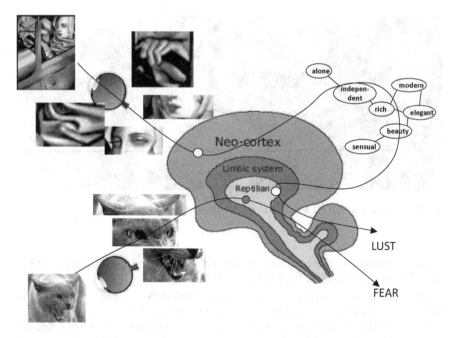

Fig. 3.5 While looking an aggressive cat, details are perceived (ears position, teeth, eyes), activating immediately *fear*; while looking the self-portrait of Tamara de Lempicka, the details are processed by the neo-cortex, building a metaphoric structure connecting: alone–independent–rich–modern–elegant–sophisticated–fine–sensual– ... activating then *lust* emotion

3. Care, Play, Panic/Grief are useful to the species without autonomous offspring, then to mammals,[8] and in particular the humans.
4. The various emotions, also hierarchic, do not impose a functional hierarchy, and interact, influencing each other (Fig. 3.6).[9]

It is evident that the capability of interpreting the cultural stereotypes of a potential user makes a designer able to decide the proper metaphors able to influence any type of emotion; more, by adding simple perceptual signals acting primarily on the brain (i.e. activating the lower levels), it is possible to reinforce the messages in a not conscious way.

The above conceptual organisation into layers, suggests that the emotions (1) and (2) are related to the reptilian brain, while the (3) are related to the old

---

[8] The concept of mammal is vague, from monotremes to whales; what is relevant is not the structure of the individual, but the non autonomous offspring, imposing care, social behaviours and so on.

[9] For example, a mammal female is unavailable for Lust during its pregnancy, and this could be the basis for the polygyny of most of the species, and for Rage for the defence of the territory; again, the alpha-male within a group protects its offspring, "using" Rage within its group; Care and Rage interacts without any predefined direction.

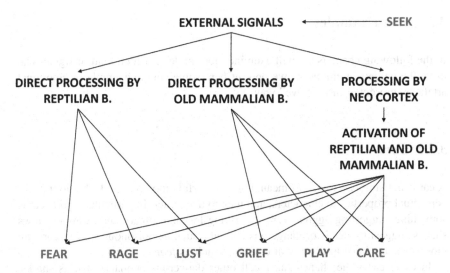

Fig. 3.6 Different emotion arousal paths according to involved brain paths

mammalian. But it is not relevant from the point of view of the consequences in our research.

According to this view, there are no positive and negative emotions: all of them are functional to the development of the species. Nevertheless, we know that we can feel pleasant and unpleasant emotions. For what I understand, the pleasure is related to the production of dopamine and endorphin, that stimulate the peripheral system, increasing our capability to react to the environment, increasing our performances useful for the evolution. Among the primary emotions, we can consider pleasant:

(a) Seeking (for any chordata).
(b) Lust (for any chordata).
(c) Care (for mammals).
(d) Play (for mammals).

But we are complex: often, the interpretation of stereotypic signals that could be unpleasant are relocated as pleasant through Play.[10]

---

[10] For instance, horror movies, thrillers, vampire literature, as well as many kid plays are perceived and searched as fiction through Play; to feel Fear consciously recognising the unreality of the danger gives pleasure.

3.2 Some Examples

In the following chapters we will examine more in detail which kind of signals can drive the perceived emotions; we present here some simple example of emotional artefacts, presented in a "naïve" way.

3.2.1 Cars

A car is not simply a transport mean, but a powerful emotional tool, able to transfer perceptual properties to characters, and then to the owner. In particular, sports car is something suggesting aggressiveness, striking fear into others: the example shows shapes suggesting heads of dangerous animals, such as poisonous snakes; for sure those artefacts have to cope with emotions such as *fear* and *rage*.

In other cases the shapes can recall other dangerous animals, such as sharks, coping with the same emotions (Fig. 3.7).

Fig. 3.7 Sports car, vipers and sharks

3.2.2 Home Appliances and Care

The house is the kingdom of the family, of the group gathered usually (in our stereotypes) around a woman; such a woman can be the sweet mother, or the strong servant, and, in some case, just the man has the control on mechanic or electronic technologies.

In the following figure, the former cheese grater, with its rounded and mild shapes, with plastic material and soft colours suggests for sure *care*; the latter, heavier, in metal, with more aggressive shapes, is more masculine and attacks the cheese with *rage*.

The first espresso machine seems more technical and "squared": again, *rage*; the second recalls a human figure, with eyes and mouth, as a servant offering a coffee: the *play* emotion rises; the third, rounded, and mild, is more related to the mother and to the *care*.

On the second row, two advertising messages on a house detergent and on a delicate washing soap: the described situation in the former is explicitly related to *play*, while the latter, through the smiling, soft, rounded puppy (mascot of the detergent), evoke babies and *care* (Fig. 3.8).[11]

Fig. 3.8 Cheese graters, espresso coffee machines, detergent advertising and mascot

---

[11] The small bear is called *Snuggle* in USA, *Robijn Robijntje* in Belgium and the Netherland, *Coccolino* in Italy; *Coccolino* derives from Italian "coccolare", "to snuggle"; more, the final inflexion "-ino", in Italian stands for small and pretty; the Belgian name recalls a precious ruby.

In all the cases, the shapes and other related perceptions are used in order to drive the user toward specific feelings, just suitable for the market target of the goods.

3.2.3 Health Care Environments

Health care involves many emotional aspects in the patients, being they in a negative psychological attitude, related to fear, uncertainty, suffering, and so on. Emotional Design cannot, of course, affect the healing process, but can largely improve the psychological status, and possibly improve the effectiveness of the therapies.[12] We mention here three done actions, some of them with measured or qualitatively verified effects.

3.2.3.1 Disguising an NMR Machine at Children Cancer Centre Pausilipon in Naples

A NMR machine is an impending object, imposing to stay motionless for a while, accepting annoying noise; its use is unpleasant for adults, and more and more for children; in order to avoid failures in the operations, the children are subject to sedatives,[13] not only calming them, but also reducing their motility. In the Cancer Centre Pausilipon in Naples, the Radiology Department Chief physician asked an artist to disguise the machine, without influencing the operational aspects: the artist covered any part of the machine with funny pictures, creating optical effects, rhythms, intriguing images.[14] The children, during the examinations, activated their *seeking*, reducing *fear* and psychological *grief*: the result has been a dramatic reduction of sedative administration (Fig. 3.9).[15]

3.2.3.2 Emotional Service Environments in Health Care

Patients spend a lot of their time in bureaucratic activities and in waiting. During both, their emotional status is very negative. Simple intervention, and often without costs, can improve dramatically the situation.

---

[12] According to some authors (Soresi 2005), studies on Psycho-Neuro-Immunology show that the emotional status of a patient can influence dramatically the healing process, as well as the effectiveness of the response to the drugs and to the therapies.

[13] The administration of sedatives affect about the 40 % of the patients.

[14] The artist was famous: Mimmo Paladino, who offered free of charge his work; the Department Chief, source of all these technical information, is doctor Enzo Salvi.

[15] About to 2 %!

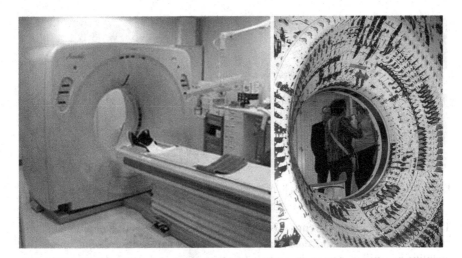

Fig. 3.9 A NMR machine, and a similar one "disguised" as a toy

The first picture refers to the reception desk of the Local Health Care Unit *ASL8*, in Asolo, a small town in Northern Italy; beside tickets for ruling the queues, large screens to warn a call and comfortable upholstered chairs, the desk has an unusual shape: while usually the reception desks are convex, as a semicircle, with many places at the circumference for coming people and few people in the central internal part, in this case the shape is inverted, and is concave toward the people: the emotional result is very interesting: while usually the clerks are perceived as inside, and the people is outside, here the contrary is perceived; more, the desk is "embracing" patients. *Inside* and *outside* are powerful primary metaphoric interpretation mechanisms (Lakoff and Johnson 1980), coping with protection, refuge, and then with *grief*[16]: the reverted concavity reduces this negative emotions; more, the "embrace" by the desk enforces *care* (Fig. 3.10).

The latter picture is the waiting room of the Breast Radiology Department at Istituto Nazionale dei Tumori (a well known Cancer Centre) in Milan. This Department has been recently refurbished, changing simply the colours of walls, ceilings and floors: the represented room, for example, has orange and warm-white walls, straw-yellow floor, and warm lighting at ceiling; more, the chairs are upholstered with deep purple fabric. In other parts of the Department, the doors purple, fuchsia, acid-yellow, green constitute a colour code unusual for hospitals, and the same for walls and floors. As the whole Department, the waiting room is filled of many paintings, making the it an intriguing set of colours and shapes.[17]

---

[16] In many of his works, Panksepp refers *grief* as *panic/loss*: he outlines how *grief* is related to the panic for abandon, as babies without the mother.

[17] An international call, done just though e-mail and "tell a friend", provided free of charge more than 170 painting by more than 160 artists over 10 countries. Many of the paintings was done properly for the departments, to communicate, in the interpretation of the artist, hope and life wish.

Fig. 3.10 A reception desk at ASL 8 in Asolo, and a waiting room at INT in Milano

The results of those intervention is a general perceivable positive emotional status crossing patients, doctors and staff. Beside some naïve observation about the quantity of smiling people in respect to other more traditional departments, non structured comments have been recorded from many patients, putting in evidence their strong *seeking* activity, positively influencing the emotional status.

3.2.4 Services

The previous examples are related to physical artefacts, i.e. to elements we can experience in the domain of *space*; similar considerations can be done also for *services*, and in particular in the domain of the *time*.

The typical Italian Immigration Office (for stay or work permits) is physically characterised by a neglected environment: uncomfortable queues driven only by rusty barriers, grey walls, dark environments, scarce presence of uniforms and aggressive aspect of policemen: those physical elements have to cope with *fear*, *rage*, *grief*, and, in some way, are functional to reduce reactions by applicants, to avoid complaining on personal rights, and so on.

But the procedural dialog, the very important part of the service, is rhetorically carried on in the time domain, with the same goals: the applicant is expected to answer to questions, not to ask for rights or solutions, often the popular form "*tu*" (familiar "*you*" in English) is used by policemen toward foreign applicants, which are forced to use instead "*lei*" (courtesy form of "*you*" in English); orders take place of explanations, and so on.

Similar behaviours can be observed in other services, and, at least in Italy, with different approached depending on the fact that the service is issued by a public or a private organisation: the procedural rhetoric used at a desk of a public hospital is often characterized by hurting communication (*grief, fear*, "I know the rules, you are nothing"), while the same desk in a private clinic is gently and smiling (*care*, against *grief*, "I am at your service, you are the customer, giving me my salary").

Everybody experienced the interaction with a call centre, and everybody had good and bad experiences. The customer satisfaction is more related to the emotional content than to the effectiveness of the service.

Example 1: Call Centre of a telephonic company[18]
"*Answer from Operator no. 127.*
– Hallo, my name is Maria. How can I help you?
– I am a your customer, and my fix telephone line does not work; my number is xxxxxxx, and I am without connection since yesterday at 10 pm.
– You are Mr. x, and your address is xxx, right?
– Right.
– Just a moment, I'm doing a technical check. Hold on
... *5 min of music...*
– Ok. Mr x, your line does not present any problem.
– But... my telephone is not working!
– Maybe is a problem of your telephone.
– No, I changed it, and also the internet line is down.
– I'm sorry, Mr. x, For me anything is Ok.
– Can you organise a visit by a technician?
– You can connect to our website, to open a ticket.
– But I have no connectivity!
– Sorry, I cannot open a ticket for you.
..."

It is absolutely evident that the experience was soaked in *grief*, absence of *care*, growing *rage*. The problem was not solved, the customer felt him/her absolutely alone.

Example 2: Call Centre of car company
"*Answer from Operator no. 234.*
– Hallo, my name is Alice. How can I help you?
– I am xxx, my car, the model xxx, started to sound the alarm loudly without any reason, and I cannot stop it. What can I do?
– Oh, I understand, it is a very embarrassing situation! Did you try to re-start ignition?
– Yes, without any effect.
– I see from your phone number that you are in Italy, and the address is xxx. If you want, I can call for you an authorized garage near to you, to help you.
– But it is Saturday night: I don't think someone will come now.
– Of course, but I can arrange for an urgent intervention on Monday.
– But I cannot leave my car sound for a couple of day!
– You are right! Oh what a bad situation. What can *we* do? Are you able to detach the battery?
– I have not the proper tools.
– Oh, I understand! I could suggest you to go with your car till the garage xxx; is not far from you! They will not solve the problem now, but will be able to detach the battery. Is your machine is running?
– Yes, but...
– Oh, I understand that it is very embarrassing to drive a howling car in your town, at this hour! but the garage is quite near... maybe, if you put on the hazard lights, your

---

[18] The example here referred are real cases.

neighbourhood will understand... I am really sorry, really, but I cannot imagine other helps
I can give you!
...”

As in the previous example, the problem was not solved, but the feeling was of participation, empathy, *care*. The customer was not alone. And the service was felt as very good!

The above example show that emotions are not only a relevant part of the experience with artefacts, but also with services.

Service Design and Interaction Design have to cope with emotions, and start to become Service Emotional Design and Emotional Interaction Design. What we present in this book is applicable to emotions both in space and in time domain. As we widely will show in the following chapters.

Chapter 4
Perception and Emotions

Abstract In this chapter, the main simple perception elements are examined, able to influence emotions. In particular, the principles discovered and organised by Vilayanur Ramachandran are examined. Many examples are provided as applications in design and communication.

Keywords Peak shift · Perceptual grouping · Contrast · Isolation · Perceptual problem solving · Symmetry · Generic viewpoint · Repetition · Balance

According to the scheme of Fig. 3.5, the signals coming from the external world are processed first by the sensorial organs and then by dedicated parts of the brain, in order to produce some kind of meaning, and then emotions.

Many studies have been carried on showing the influence of the shapes on the mood of the observers, and many others have been carried on observing the emotional reactions to the colours.

Among the former, it is well know the *bouba/kiki* effect: initially observed by the psychologist Wolfgang Köhler (1929), has been more recently re-examined by Ramachandran and Hubbard (2001); asking to a number of people which of the two shapes (Fig. 4.1) is called *kiki* and which one *bouba*, more than 95 % of the answers attribute *kiki* to the former and *bouba* to the latter, and the experiment has been proved independent by cultural or linguistic aspects.

The experiment shows that there are synaesthetic relationships between the sound of the words and the shapes, but put also in evidence as edges are perceived as aggressive shapes, and related to hard sounds, while round shapes are related to smooth lines and sounds. Further experiments shown that it is possible to measure the emotions related to specific shape characteristics (Lu et al. 2012).

Also the literature on the colours and their emotional effect is rich (Whitfield; Whiltshire 1990), and in particular for marketing aspects (Aslam 2006).

There are no studies relating perceptual aspects to the influence on the Panksepp's emotions, but very interesting works have been carried on by Ramachandran, on aspects we can easily connect to Seeking and other primary emotional status (Ramachandran and Hirstein 1999). We will refer here just to those studies,

Fig. 4.1 Which shape is called *kiki* and which one *bouba*?

providing examples in the field of design, showing the properties an artefact should present in order to attract the attention and to raise emotions.

4.1 The Ramachandran Principles

Ramachandran presents nine[1] perceptual characteristics that he verified, mainly through functional magnetic resonance, able to increase the production of some neurotransmitter (mainly dopamine) and to rise the corresponding emotions.

In the following, we examine in detail each of them, providing examples.

4.1.1 Peak Shift

Peak shift means exaggeration of some aspects against the balance of reality: the choice of the enhanced aspects extracts the "truth" from many contingencies; in some way, physics extracts meanings from reality by "peak shifting" in the experiments: dynamics describes referring to mass, forces, acceleration, regardless shapes or colours of the involved objects.

The ability of the brain to recognise peak shift is well expressed by some laboratory experiments with rats: training rats to react to rectangles in a different way from squares, it has been observed that the animals respond more strongly to those rectangles in which the ratios between the sizes are more exaggerated.

The experiments show a certain capability of abstraction, but also of the capability to choose what to abstract; from a neurophysiological point of view, it could be put into a relationship with the need for a prey (or for a predator) to recognise quickly distinctive elements of predators (or preys), independently from details. This capability should be cross-modal, involving sight, smell, sound, etc.

In Fig. 4.2 there are examples of artefacts presenting peak shift.

---

[1] In some popularizing books, he adds a tenth principle, the metaphor, but this is not examined in his scientific production. We consider metaphors more related to the complex perception, dominated by the neo-cortex, and that will be examined in the following chapters.

Fig. 4.2 Examples of peak shift: armchair *Proust* by Mendini, a bear puppet, a banana-tray, a lemon squash by Alessi, a spaghetti cooking fork, the Neolithic *Venus of Willendorf* (drawing by the author), *Nueschwanstein Castle* in Bavaria, Germany (it was used as inspiration for Disneyland's Sleeping Beauty Castle)

The first is the armchair *Proust* by Mendini for Magis: its heavy structure is the caricature of an old rich armchair of the XVIII century; it is made of polyethylene, often in vivid colours, in contrast with the richness of the shape and of the ornaments, and this fact increases the attention, and then seeking, on the peak shifted shape, unusual for the material.

The second is a plush bear by Les Mills: the exaggerated sizes of head and pawns recall immediately a puppy, inducing an automatic arousal of Care.

The third, a banana-tray by Fackelman: the peak shifted shape enforces seeking.

Mysqueeze by Kreiter and *Juicy Salif* by Starck, both for Alessi, emphasize the squashing heads, deleting as much as possible further details; the peak shift operation result in the exaggeration of the only functional element, suspended in the space, in name of Seeking, no matter about the operability.

Then the service fork *Tibidabo* by Lassus for Alessi: the number of tines exaggerates the concept of fork, increasing immediately the attention on this aspect.

Then, a Neolithic sculpture, known as *Venus of Willendorf*, emphasizing the maternal attributes and shapes.[2]

Finally, the *Nueschwanstein Castle* in Bavaria, Germany: not a fairy tale castle, but a real one, built as an the exaggeration of any possible fairy tale castle. It was commissioned by Ludwig II of Bavaria as a retreat and as an homage to Richard Wagner.

4.1.2 Perceptual Grouping and Binding

The human brain tends to group and bind phenomena in order to gather them around a unique already known explanation. For instance, many optical illusions work in this way, and in particular the well known image of the Dalmatian, got through a high contrast picture (Fig. 4.3).

The principles described by Ramachandran correspond largely to the law of the Gestalt, and to the rules largely described by Kanizsa (1997).

Used mainly in graphic design, those elements are applied also to interaction design, and in some cases also to product design. For example, in Fig. 4.4, we can observe the *9093 Kettel* by Graves for Alessi, in which the handle is a perceived as a perfect circle, with a ball in the centre, and the wall clock *DIY* by BOX32 design for Karlsson, in which separated elements are composed so that to rebuild a unique well recognised composition.

The Ramachandran principles are largely used in graphic design, and constitute a valid support to drive the attention in web design. In Fig. 4.5 the homepage of the architectural group www.big.dk is presented, in which each small square corresponds to a project; the different colours of the squares correspond to different kinds

---

[2] The sculpture is part of the collection of Naturhistorisches museum in Wien. By the way, Ramachandran's examples on peak shift are referred to the Indian sculptures of the goddess Parvati, emphasizing the feminine elegant and pretty body.

Fig. 4.3 A well known picture: a set of black spots, in which is it possible to recognise a Dalmatian; the recognition, due to the proper grouping of the spots, with a pleasant effect, can be put in relation with the activation of the limbic system

Fig. 4.4 The *9093 Kettel*–Alessi and the wall clock *DIY*–Karlsson

of projects, and the perceptual grouping gives us a lot of information about the activity of the company; moreover, each square is "peak shifting" the shape of the project, that can be revealed by a thumb image when rolling over, or access for details by clicking the mouse.

Fig. 4.5 www.big.dk (April 2013), the home page and a detail when rolling over the mouse

4.1.3 Contrast

Rods and cones in the human retina are organised to emphasize the perception of edges and contours; other parts of the brain in the visual cortex respond mainly to such an extracted information; so, a line drawing or a cartoon stimulate these cells as effectively as a 'half tone' photograph. Such contrast extractions seems to be related to "pleasing to the eye". According to information theory, information exists in the regions of change, e.g. edges, and then such regions get more interest by perception. For sure the contrast areas are informationally more interesting, physiologically more perceivable, emotionally more meaningful. We can say more pleasant.

Contrast is for sure related to the perception of shapes distinguishing between background and foreground, and mimicry, as negation of contrast, is often a strategy used by prays for saving themselves from predators.

Contrast can be used then to emphasize some aspects of a product, we want to focalize; for example, in Fig. 4.6 the projector *P1* by Asus, present a double contrast between the lens barrel and the body: silver colour against black and round shape against squared. Those contrasts make strongly recognisable the lens, and put strong attention to it, as responsible of the quality of the product.

In web design, it is very interesting the use of contrast in the home page of Google (Fig. 4.7): the coloured logo with a search window on a completely blank background not only constitutes a strong contrast, but also points out to the seeking activity of the search, through perceptual seeking arousal.

4.1.4 Isolation

This principle refers to the isolation of a single perception modality before amplifying the signal in that modality.

For example the portrait of Ghandi (Fig. 4.8) remain always recognizable without colours, or without half tones, or with more destructive manipulations, due to the fact that the contour and the lines are privileged in our visual perceptions;

Fig. 4.6 *P1* LED projector by Asus

Fig. 4.7 The home page of the Google Search Engine

contours are the signals that we isolate in order to recognise the person; Ramachandran provides the example of the *Icarus* by Matisse (a black shape on a blue background) also in that case the *form is isolated from the background before to give it a meaning* (both in the construction and in the perception). Isolating a single area allows one to direct attention more effectively to this one source of information, having notice of the enhancements introduced by the artist, and then amplifying the limbic activation and reinforcement produced by those enhancements.

Fig. 4.8 A *black and white* portrait of Ghandi, and some manipulations loosing information, but allowing recognisability

Studies by Ramachandran on autistic subjects show their exceptional drawing capability, as a result of the isolation on contours they apply in observing the real world. It is the case of the subject Nadia, a 5 years little girl, drawing with exceptional capability horses and other animals, with a quality comparable to the one of famous masters.

Isolation works mainly on the shapes, and there are many design masterpieces that owe their recognisability just to this principle (see Fig. 4.9).

Isolation is possibly one of reasons of the success of many mass consumer products. For example, the shape of Coca Cola bottle allows its use as a product symbol with high recognisability; the same for the shape of the Absolute vodka; the Campari Soda bottle has been for the same reason used to build lamps (Fig. 4.10).

4.1.5 Perceptual Problem Solving

Ramachandran argues that to discover a signal in noise is rewarding in itself; as well as in Dalmatian picture we build a meaning from spots, getting a form of "pleasure", in the same way we can get emotion in finding a solution from clues; not only in seeing (the Dalmatian example; also, more sophisticatedly, erotism looking a nude through a veil instead of the image of a woman nude in the open; again, some cubist or surrealist picture, as many Dalì's paintings like *Slave Market with the Disappearing bust of Voltaire*, or the many "bi-stable" figures, as the ones shown in Fig. 4.11).

The designer can exploit this principle both creating objects not evident in the use, but comfortable after having understood them, or simply surprising in their properties; sometimes mimicry can be a weak expression of perceptual problem solving.

In Fig. 4.12 many examples are presented.

Fig. 4.9 The *Red and Blue Chair* by Gerrit Rietveld owes its strong recognisability to the straight lines and the geometrical shape, that prevail on other aspects; the lamp *Atollo* by Vico Magistretti has a unique profile, and we isolate the unique contact point between the basis and the hemispherical shade

The first is the ergonomic seat *Varier*: the first time one sees it cannot imagine its use, and only after a trial can appreciate the ergonomy and the comfort.

The Tube Chair by Joe Colombo is beyond any suspicion, in particular if observed when extracted by its packaging, being a set of tubes of different diameters one inside the other.

The lamp *As much you need* by the Korean designer Hong-kue Lee, as most of the touch-commanded objects do not present significant clues for its use: by extending the hand on the top of the lamp, and sliding a finger in correspondence with an array of LEDs, the user turns on as much LEDs as the extension of the hand. After having discovered the clue, the customers are strongly satisfied and emotionally happy.

4.1.6 Symmetry

Not the symmetry, but discovering it seems to be rewarding: according to Ramachandran, symmetry is a typical aspect of the predators, and our eyes and brain are properly organised to detect it; but symmetry is also economy in knowledge, because the property of a part can be used to understand a whole. More, symmetries can be transposed from visual aspects to many others, and it seems that the neural

Fig. 4.10 The power of the isolation principle: the recognisability: two bottles of Coca Cola (25 cl and 33 cl), with the same shape and different proportions; the silhouette of Absolute vodka (a large advertising campaign has been based on the bottle shape recognisability, often as an empty space built through other objects drawing its contour; the bottle of Campari soda: based on its bottle personality, Ingo Maurer designed many chandeliers and lights built just assembling bottles (by the way, also the red tonality of Campari soda is very well recognisable, and those lamps maintaining such a colour)

centres activated to discover the visual elements are the same working with the many other perceptual channels.

It is quite difficult to find asymmetric design artefacts, and maybe it is due to the fact that we as human have bilateral symmetry: only ergonomic aspects (e.g. handles or knives) taking into account our asymmetries (e.g. left and right hand), or functional aspects (e.g. the blade of a knife) present asymmetric shapes; sometimes, the asymmetry is used just as a natural violation, as attraction point (Fig. 4.13).

Fig. 4.11 Three well know "bi-stable" images: each of them can be interpreted in two different ways, but while looking at an interpretation, the other one is not perceivable; the observer satisfaction in discovering the clue persists after the first observation, and can be repeated many times: *All is Vanity* by C. Allan Gilbert; *Saxophonist-Woman face* used by the cognitive psychologist Craig Mooney; *My Wife and My Mother-In-Law*, by William Ely Hill (1887–1962), Puck, 1915

4.1.7 Generic ViewPoint

The human visual system is a Bayesian deduction machine (roughly, we take information from experiences, and re-apply them till that allow us to make the best predictions). This implies that when we have many possible interpretations of a same scene, we will choose the most likely.

Starting from this remark, Ramachandran and Hirstein introduce two artistic perception principles: we prefer generic viewpoints, and we abhor coincidence.

While seeking, a person behave as much automatically as possible, supposing that the perceptual signals can be interpreted according to the past experiences. Problems arise when the interpretation results wrong or difficult for ambiguities: in this case, the raising emotions are negative, bothering the user. What designers call *affordance* can possibly be considered as a hint for using generic point of views.

There are many examples of irritating violations of this principle, for example in opening-closing actions.

In Fig. 4.14, two identical common symbolic codes have different meanings the circle indicating that the switch is off (the flow circuit is open) is used in air nozzles of many cars as "air flow is on" (the flow circuit is closed); more, handles with different orientation of the edges should, in the mind of the designer, suggest when pull or push the door for opening, but the code is not intuitive; radial symmetry in

Fig. 4.12 On the *left*: the ergonomic seat *Varier*, the *Tube Chair* by Joe Colombo and the touch table lamp *As much as you need*; on the *right*, *Equilibrium* bookcase by Malagana Design, the corkscrews *Parrot* and *Anna G.* by Alessandro Mendini for Alessi and *Juicy Salif* by Philip Starck

doorknobs suggests (in the order): push or pull, do not rotate (but it is not clear which of the two actions), push on top and then rotate or push or pull the door, push on the top and then push or pull the door, rotate as a traditional handle (while the user has to push or pull the black part, according to the opening direction of the door; finally, the panic bar in a door without the information of the position of the hinge can reduce the opening capability.

Fig. 4.13 Most of the design artefacts present bilateral symmetry: chairs (in the figure, *Masters* by Starck for Kartell), cars (in the figure, an Alfa Romeo concept), and most of the examples provided in this book; some asymmetries can be found only for natural asymmetries in the users (Yamaha *RGX121Z* guitars *right handed* and *left handed* players), or for functional aspects (the blade of the knife *Piemont* of Villeroy Boch); sometimes, metaphoric recalls violate symmetries, carrying to perceptual problem solving or to other phenomena, inducing Play emotion (in the figure, the cutlery *Bite* by M.A. Reigelman)

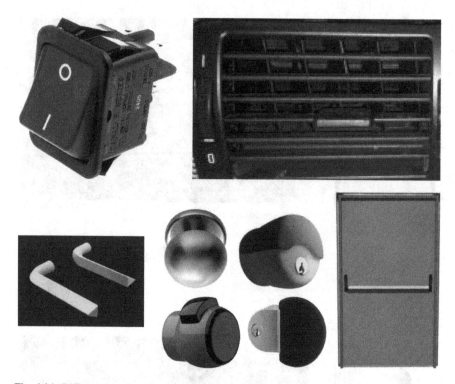

Fig. 4.14 Different meanings for the same symbols in electric circuits and in ventilation circuits; different and not always easy interpretations of the use of handles, due to different shapes

4.1.8 Repetition, Rhythm, Orderliness

Ramachandran does not give explicit examples of this aspect, but we can easily extend the ideas previously discussed: patterns, in a broad sense, induce the perception of regularities, that reduce the effort for recognition; the hypothesis that our brain is specifically organized to recognize regularities corresponds to the idea that repetition, rhythm and orderliness are the basis for a learning process.

Many examples of textures and patterns are present in arts, from complex Arabic tiling, to the drum rhythms, to more complex musical structures. In Fig. 4.15, we provide just few simple examples.

4.1.9 Balance

Also in this case, there are no explanations nor examples related to this principle, in Ramachandran's works. Being the visual perception his field of interest, we can induce the validity of this principle in a distribution of visual "weights", supposing

Fig. 4.15 *Rhythm 2152*, pendant by A. Levy, the lamp *Falkland* by B. Munari, *Two Speeds* stairs by the Korean designer Youngju Oh (the steps are higher twice than the usual, but their disposition allow to step down as in any other usual stair), the low cost polycarbonate chandelier *Namu* by Y. Oh, the surface of the paper giant chandelier *Madrepora* by Y. Oh, set of modular multipurpose poufs *Fortyfor2* by Y. Oh. All those artefacts present rhythm, repetition and order, at different scales: in the whole shape (*Rhythm* and *Falkland*), in the structure (*Two Speeds* and *Namu*), in the surface texture (*Madrepora*), and in the modularity of the possible compositions (*Fortyfor2*)

an unbalanced view neurologically able to draw attention to potential dangerous situations. The feelings related to this attribute have been examined pointedly and elegantly by Wassily Kandinsky (1947).

Chapter 5
Metaphors and Design

Abstract The Neo Cortex role is examined, and the concept of metaphor is introduced, from the point of view of Cognitive Sciences. The concept of stereotype is discussed. A simplified formalism of the semantic networks is introduced for representing knowledge and describing/discovering metaphors. Many examples in design and communication are presented and discussed.

Keywords Metaphor · Stereotype · Semantic network

5.1 The Neo Cortex

In the previous chapters we discovered that:

- there are two "inner" brain layers responsible of the emotions;
- there are peripheral systems cooperating-antagonist, independent-dependent from the emotions coming from the central system.

Some of the emotions originated by Seeking can be started by the interpretation of signals kept through perceptual channels: for example, the Gestalt rules (Kanisza 1997), as well as the principles described by Ramachandran can raise pleasure or disgust, or, in any case, bias the interpretation. But the formal elements described by such studies, despite useful, are too weak for setting a method for Emotional Design.

We need to approach the functions of the Neo Cortex. This is the more complex part of the study, but, unfortunately, it is also the part more responsible of the emotions communicated through artefacts.

The complexity is increased by the many interactions with the other parts (central or peripheral) of the nervous system. An example of the interrelationships among the various levels can be observed, for instance, in the masculine behaviour of a male-manager in a company: the Rage of a reptile "used" for protecting its safety, went evolutionary transformed into territory protection for a jackal, then into the dominance Rage of the alpha-male in a baboon troop, then into the alpha-role of a male in a human family, then in the same role of a manager in a company, then in

© The Author(s) 2015

M. Maiocchi, *The Neuroscientific Basis of Successful Design*,

PoliMI SpringerBriefs, DOI 10.1007/978-3-319-02801-9_5

the use of many stereotypical symbols (dressing code, car type, gestures, etc.) enforcing the communication of the dominance; in this chain, the first elements are typical of the reptilian brain, while the last are mainly a product of learned codes, interpreted by the cortex.

More, the same basic emotion can be transformed into leverage for behaviours, giving the chance of having many nuances of the same emotion, and many different names (explaining the couple emotion-situation). For instance, the Panic/Grief properly intended as suffering by abandoned non autonomous offspring (related to the Fear of this event) is turned into what we used call "shame" in the societies in which the group is very strong; socially unacceptable behaviours are prevented by learned rules and the sense of "shame", inducing the Panic/Grief coming by the disapproval of the group (as an extension/abstraction of the parents). Possibly, all the "secondary emotions" (as shame) can be partially embodied in the inner brain levels, and partially learned in the neo cortex. We consider, as working hypothesis, that all the secondary emotions can be led back to the seven primary Pansepp's emotions.[1]

5.2 Stereotypes and Metaphors

Our brain builds stereotypes relating the emotions to perceivable elements. Gestures, postures and expressions can be connected to basic emotions, as well as to secondary emotions (Darwin 1872; Ekman 2003). So we are able to interpret the emotional status of a person with Rage or Fear, and we recognise supposed or real leadership, shame, glory anxiety, pride, and so on. This recognition can be related to static signals (e.g. dress code) as well as to scripts (stereotypical stories, with role playing, causal and/or temporal dependencies, etc.). This is what Lakoff calls *metaphor* (Lakoff and Johnson 1980).[2]

Proceeding from the inner brain levels to the cortex:

- many metaphors are the basis of our behaviour and though (in vs. out, from-to model, etc.); this kind of primary metaphors are common to humans and many animals;
- many metaphors are organised as stereotyped scripts ("you Rage, I dead, I am not interesting for you" is possibly the basis for thanatosis; other scripts survive into proverbs such as "Those who make themselves sheep will be eaten by the

---

[1] In this sense can be interpreted the Plutchik's wheel of emotions (Plutchick 2002).

[2] According to Lakoff, a metaphor is an operation based on two different semantic fields, implicitly modeled as entities and relationships; the recognition that such two fields have isomorphic models allows us to refer semantically to one field by using verbal references to the other. For instance, if we recognise that the structure of nourishment (a morsel enters in the mouth, goes in the stomach, is then digested, and provides vital energy) is similar to the one of the communication (a word is entering in the ears, goes to the brain, is then understood, providing knowledge), we can jump between the two semantic fields with phrases such as "I have not digested your words" or "what you are saying is honey for my ears".

wolf", "His bark is worse than his bite", "Mors tua vita mea"): those metaphors are common to humans and many animals;

- other metaphors are structured into myths (f.i. the Greek mythology), religions, rites; for what we know, those are peculiar of humans;
- finally, other metaphors are organised through progressing experiences, with a collective exchange, diffusion and endorsement, but they conduct always to the primary emotion centres[3]; the association James Bond—Aston Martin includes many steps, each of them conscious and rational (adventure, strong, hero, alpha male, sport, sophisticated, beauty, women, rich, and so on), but all of them are filtered by cultural aspects, reducing the metaphor to very few elements, conducting to Seeking, Rage and Lust.

5.2.1 Basic Metaphors

Simple paradigms as in-out, from-to, up-down are the metaphoric basis for the interpretation of the world and the interaction with it.

For example, in-out refers to the relationships between container and content, and we can relate this paradigm to many objects and actions: to put something in a box, to eat something, to live in a house, to collect objects and so on. By the way, a collection of objects uses the paradigm in–out in a more abstract way, being the notion of "in" related to some kind of classification and to the abstract operation to decide whether or not an element belongs to a class. According to Lakoff, these paradigms are embodied in our physical brain structure, and are related to Seeking.

While designing any kind of artefact, those basic metaphors are often used without any kind of consciousness. For example, the designer knows very well that we sit *on* a chair, but sit *into* an armchair: so, an armchair should kind be a container, and the goal of the design will be the qualification of such a container, adding properties able to suggest further meanings. So, the armchairs shown in Fig. 5.1, present different meanings, but all of them induce the idea of a container.

The presence of these simple metaphors also in many animals is evident. For example, the model inside-outside is the basis of the construction of tools implemented to catch water, to extract insects from holes, and of collecting objects (Fig. 5.2). While these metaphors induce to think to some abstraction activity performed by the brain of the animals, others are absolutely instinctive, embodied in the animals, such as to build a nest or to dig a lair, or to hide bones as dogs do (once more inside-outside).

Beside the conceptual role of the metaphors, such phenomena are allowed by the capability of primitive characteristics of our sensorial channels while Seeking. For instance, our eyes (whose model we usually simplify as a camera) are complex tools

---

[3] This fact has been proven on laboratory mice; the neocortex is not involved at all in the primary emotions (private communication with J. Panksepp).

Fig. 5.1 Different armchairs, each of them suggesting the concept of sitting into. Other metaphoric signals are added to each of them, changing completely the "meaning". From *left* to *right*, rich and powerful, classic and modern, caring of you, a lair, young and ironic. The added properties make references of other metaphoric signal, processed by other brain levels, as we will discuss further

Fig. 5.2 A bonobo using a stick for fishing termites, a crow using a stick for extracting insects from a tree, a satin bowerbird collecting blue objects in its nest; those are just few examples of animals building tool on the basis of inside-outside model: some apes build glasses for catching water, magpies and raccoons collect shiny things, and so on

able to perceive a representation of the external world, enhancing through some "biological trick" the contours of the different objects,[4] and the collected signals are then processed by some parts of the brain, able to gather separate signals to build a complete image[5] (Fig. 5.3).

This capability is the basis also for the recognition of common artefacts, such as faces; in particular, we have parts of our brain devoted to the specific action to recognise a human face (Zeki 2009)[6] (Fig. 5.4).

---

[4] This capability allows us a proper recognition of a figure on a background; more it is used by many animals for mimicry (for example, the white fur of a rabbit on the snow deletes any contrast and contour) (Pierantoni 1996).

[5] The phenomenon is well described by the Gestalt laws and the studies of Kanizsa (1997).

[6] By the way, this characteristic is present for the chicken (Vallortigara 2005).

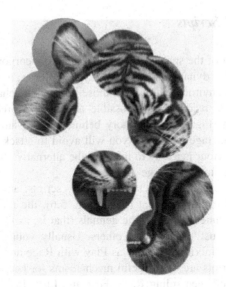

Fig. 5.3 Different visual signals coming simultaneously to our senses are locally interpreted, then integrated by our brain, so that we can recognise a whole just looking some small parts, as the tiger in the example

Fig. 5.4 Our brain gather signals that can be interpreted as faces

5.2.2 Embedded Scripts

During the evolution of the species, the brain structure incorporates not only static interpretation, but also dynamic scripts, prototypical stories, as models of possible interactions with the environment. One of those instinctive behaviours is related to thanatosis, that is the reaction to a possible attach by a predator consisting in simulating the death (Fig. 5.5). The story behind is "if I am dead, you are not interested in me, your rage is useless, you will avoid to attack me".

This kind of behaviour is a way to interpret the alternative fight or flight, and is for sure relate to the emotions Rage and Fear.

Other behaviours can be conducted to primary scripts: animals presenting the belly and the throat to an opponent (Fear) (Fig. 5.6), the mimicry in mandrills, whose face has the same colours of the genitals (that is, as to say "do not attack, leave Rage, accept Lust"), and many others. Usually young animals use those scripts while playing, mixing emotions as Play with Rage and Fear (Fig. 5.7).

These primary scripts are often useful mechanisms for keeping "under control" social behaviours, and then ruling Rage, Fear and Lust. In Fig. 5.8 a scheme is presented, referring related to a baboon troop (Sapolsky 2002): an α male (A in the figure) is ruling the group, after he got the power through violence; he gets respect from any member through aggressive behaviours; he takes care only of his sons; he accepts grooming by his partner (B), but he is in any case dominant; the female B

Fig. 5.5 Examples of thanatosis by a snake, an opossum, a kingfisher and a frog

Fig. 5.6 Wolves show belly and throat to accept the opponent as winner, saving the life; the same ritualized behaviour can be observed in plays (figure also for humans, as well represented in the movie "War of the buttons" by Y. Robert)

Fig. 5.7 Mandrills combat rage of possible opponents by camouflaging the face with the same *red* and *blue colours* of the genitals, stimulating the substitution of rage with lust

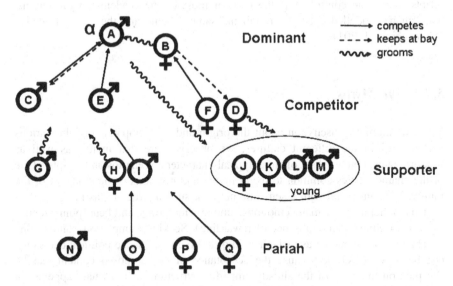

Fig. 5.8 The scheme refers to the structure and the behaviours of a baboon troop, as well as to the life of a human group into an academic environment

benefits from her social position; lower level competitors, male or female, try to substitute the α male and his partner B. A further level of supporters knows that they have no chance to compete; finally a group of outcast (possibly for physical

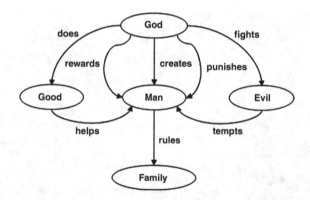

Fig. 5.9 The foundation of a family in many religions: God created the Man, and operates the Good, fighting the Evil; God rewards the Man operating the Good and punishes the Man following the Evil; the Man build and rule a Family; when putting the word Father in the place of God, we can see that the Man on his turn, will behave as God do, toward his Sons

problems) conclude the hierarchy. A rule: any level is ill-treated by any individual of any higher level. The same behaviours appear in human organisations, in particular were the absence of mandatory needs or of firing risks allows primordial scripts to get the control: it is the case of most of the academic organisations, that can be modelled exactly through the same scheme and the same structures (Maiocchi et al. 2011).

5.2.3 The Myths

Without entering a discussion about the origin and the comparison of the various mythologies in the different cultures, we observe here that myths, as well as folkloric tales, contain a set of stereotypical characters, scenarios and scripts.[7] One of the characteristics common to many myths and religions is the structure of the family. We can model with a semantic network such a structure, as in Fig. 5.9.

This model involves many emotions, among which Rage and Fear (punishment), Care (rewarding), Panic (absence of rewarding), Seeking (temptation), and so on.

The use of this model as metaphor, in order to communicate political decisions, can be very powerful, because the acceptance of the proposed content can be accepted on the basis of the already embodied structure; this is what happened in the political discourses by G.W. Bush after his election in 2001: using the above

---

[7] In particular, it is hard to understand whether or not the related models are learned or are embodied; in any case those stereotypical structures have been spread among the population since some tens of thousands of years during the ran oh the homo sapiens from the Rift Valley to any part of the world (Cavalli Sforza 2008).

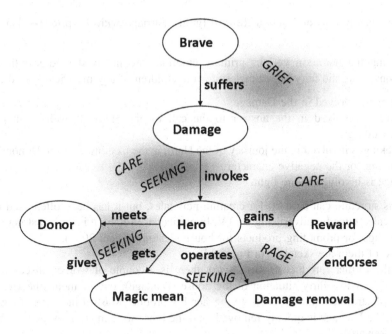

Fig. 5.10 The essential structure of a Russian popular tale, according to Propp

conceptual frame, introduced the nations as persons, and then the United States as a person willing the Good and fighting the Evil, and then introduced the notion of "Rogue State" as personification of the Evil: so the preventive war against Iraq was fitting as a right fight against the Evil, without any concrete reason for an attack (Lakoff 2004).

Other very powerful stereotypical scripts are the popular tales. In the twenties Vladimir Propp studied a specific class of Russian fairy tales (whose structure is spread overall in Europe, and not only), providing a formal structure common to all of them (Propp 1928).

Again, using semantic networks, the typical essential structure of a tale according to Propp is the one represented in the Fig. 5.10.

As shown, the story is structured in the following way:

- a Brave character (a king, a princess, a father, and so on) suffers for a Damage (a kidnapping, a theft, etc.) by a negative character (a dragon, a baba yaga, etc.);
- then, the Brave issues a call for some Hero (usually male) helping him/her;
- a Hero comes and accept to fight the negative character;
- he leaves, meets a Donor (a dwarf, an animal, a very old woman, etc.); the Donor submits the Hero to some trials, and, after passed, gives him a Magic mean (a ring of invisibility, a flying horse, etc.);
- the Hero goes against the negative character, fights him, wins, removes the Damage;

- the Hero returns and gets a Reward (gold. marriage with the princess, lands, etc.).

During the narration various primary emotions are involved (and just those emotions make the fairy tales timeless for the children of any historical period):

- Grief is involved in the Damage;
- Care is involved in the answer to the call by the Hero, as well as in his Rewarding;
- Seeking is involved in the journey of the Hero, in the meeting with the Donor, in looking for the negative character;
- Rage is involved in the fight.

This structure can be used for creating new tales, but it has been often used in many video clips for advertising on TV. In particular, in Italy, it is typical structure of the clips for promoting products for house cleaning (Diani 2014).

Two of them are sketched in Fig. 5.11.

In the former (product *MastroLindo*), there is a couple of women discussing about the terrible dirty situation of the house (Damage); one of them (the Hero) meets Mastrolindo (the Donor), giving her his product MastroLindo (the Magic mean); the Damage is quickly removed, and the other woman strongly praises the Hero (Reward).

In the latter, the tale is more evident, and the clip is done through an animation. In a castle, a king is suffering because a very dirty cauldron, that nobody is able to make shining (Damage); the king invokes some solutions (call); suddenly, through a window, a knight in armour (the Hero) enters, flying on a horse; he (at this moment) has in his hand *Cif* (a cleaning product) (the magic mean); he quickly cleans the cauldron, that becomes sparkling shiny (Damage removal); after that, the knight takes the helmet, and a woman appears: the Hero in the house is a woman; finally, the woman, happy, sits on the throne (Reward): the woman became the Queen of the house.

Fig. 5.11 Some frames taken from two Italian ads on house cleaning products; the structure of the clips is modelled according to fairy tales

The capability of recalling, ironically, the structure of the tales and the related emotions is a powerful mean to make some messages pass easily and quickly: those products are magic means, removing any dirty. This structure can be recognised in many others clips related to the same kind of products.

5.2.4 Neo Cortex Metaphors

The various levels of our brain are just a simplified model; in fact we have an unbelievable number of interconnections among the neurons, and there is no continuity solutions between the various levels; the different parts of the brain works in parallel, and with many retrofitting cycles (this is unavoidable in any kind of lattice structure); possibly, with a growing complexity, some kind of speciali-sation arises, useful to a specific culture; if we would study the different the evo-lution of the cultural values inside any population (Hofstede 2010), we could discover that stereotypes and metaphors are coherent with their cultural history, and with typical customs.

The neo cortex seems to be just the "repository" of cultural and experiential aspects[8]: learned models and information, results of personal experiences, personal mental processes and thoughts shape day by day the brain networks, increasing our capabilities, but any process, beside conscious and rational, has to cope with the emotions, that it, any process occurring in the neo cortex interacts with the lower levels, exciting the centres related to the basic emotions; on the reverse, any process coming from the farthest body part is able to reach the central nervous system, interacting with primary emotions or with cognitive capabilities, inducing emotions.[9]

The stereotypical structures seen in the tales or in the myths are very old and deep, but simpler in respect the ones we can recognise as product of the experience. For example, if we look to the two pictures below, we have a lot of similar signals, but different emotions rise; we can examine Fig. 3.4.

According to a widely spread common sense, we could model the signals net-work in two different semantic nets (Fig. 5.12).

While looking the former picture, we recognise some motorcyclists with well identified motorcycle suites; they have typically shaved heads, appear strong, have tattoos, possibly with Nazis symbols, the suites are black, possibly in leather; all those elements suggest the concept of a motorcycle gang, and the common asso-ciations to that refer to free sex, to drunkenness from bear, the use of drugs, violence, weapons: the conclusion is danger, and then Fear.

---

[8] The deletion of the neo cortex in laboratory rats does not affect their behaviours: they seems more active that their "normal" companions, as the absence of an experiential learning reduce some kind of censorship ruling the actions (private communication with J. Panksepp).

[9] For example, a pain in some part of the body, as broken leg or a pang in the heart, causes Fear or Grief, due to the processes performed by our conscious part, evaluating possible risks or limiting consequences.

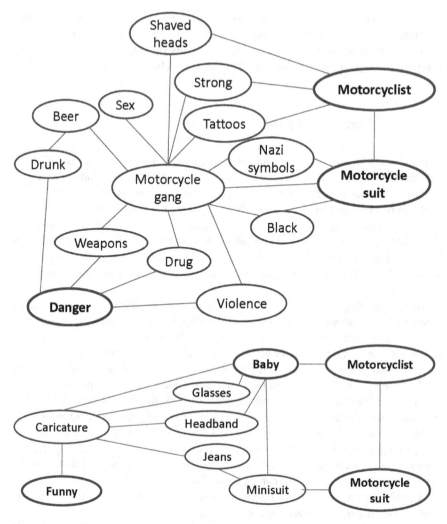

Fig. 5.12 Two different semantic networks, for two different metaphoric interpretation of the pictures in Fig. 3.4

The second picture presents a similar subject, but we recognise, beside the motorcyclist and the suite, the child: in this context, the glasses, the jeans suite, the head band recall a caricature, and then we consider the occurrence as funny, inducing Play or Care.

A further more complex example could be provided by the character of James Bond, in particular as presented in the movies played by Sean Connery (Fig. 5.13).

The elements we can gather from the posters presenting any of the movies of the series are: an appealing man, with a straight, determined confident look, often dressing an elegant tuxedo, smiling with a gun, many beautiful scantily clad

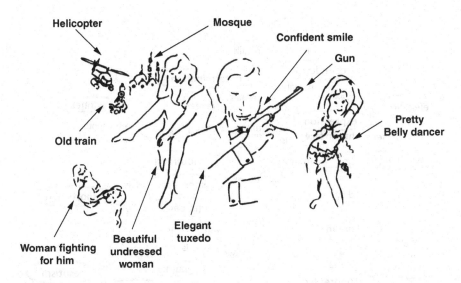

Fig. 5.13 The structure of the poster of the movie *From Russia with Love*, James Bond 007, starring Sean Connery (drawing by the author)

women, fighting women, sport cars, Turkish mosques and other exotic buildings, helicopters and so on.

Most of the posters represent a man at the centre and many details around him; i.e., metaphorically, he is the centre, and everything is moving around him. In particular, following the posters of the two movies *From Russia with Love* (presented above) and *Never Say Never Again*, we can observe: the man expresses with his face confidence and determination, without any fear (he is smiling) and with a certain challenging attitude: the man is for sure really appealing. Around him, many beautiful women, scantily clad, some of them competing for him: the man is a sexual goal. He has to cope with very unusual vehicles: sport cars and racing motorcycles are usual, but he is using also a jet pack: he is an athletic sportsman. He has a gun, ready to kill, dangerous: he is a winner. He is able to face also very dangerous, fighting with divers and with sharks: he is brave.

All those associations point out that the man is a champion among the males: within a group of animals, he would be for sure the most required reproductive exemplar. The resulting emotion has for sure to cope with Lust. Fights and the gun suggests also Rage.

The semantic network[10] that we can build in order to relate all those items are (Fig. 5.14):

---

[10] We use the formal representation of a semantic network in an intuitive way, without a rigorous attitude; for example, we use the name of the relationships in an evocative way, so that they can be duplicated.

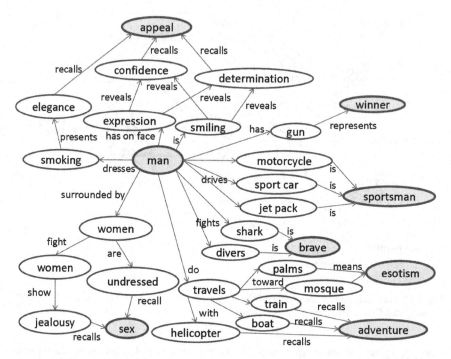

Fig. 5.14 The network representing the elements appearing in most of James Bond posters, and the possible implicit relationships (the analysis has been carried on using the two original posters of the movies *From Russia with Love* and *Never Say Never Again*)

More, his life carries him to many worldwide environments, exotic for western people, and with any kind of transportation: large boats, old trains, and so on. Seeking is stimulated.

Those emotions are pivoted on the character of James Bond, so that any product used by him will be naturally associated with that. The Rolex watches, the Aston Martin, become symbols of the reproductive adventurous desirable male, and, following the success of the movies, acquire those representation capability also out from the James Bond's context (Figs. 5.15, 5.16 and 5.17).

Of course, each metaphoric structure can be more or less acceptable according to different cultures: for example, Bond is strongly coherent with an Anglo-Saxon culture: a judging God, strong, avenger, righteous, but is also "masculine" and "individualistic", and could be not compliant with a "feminine" or a "collectivistic" culture.[11] Further studies could show that Rage and Lust prevail in the metaphors for people passed from the status of hunter-gatherers to conquering nomads, and

---

[11] The terms "masculine" and "feminine" are intended as Hofstede intends (Hofstede 2010), that is societies that strongly differentiate the roles according the genders or societies that do not differentiate them; similarly, "individualistic", learning and using "I" as a primary values, versus "collectivistic" using "we".

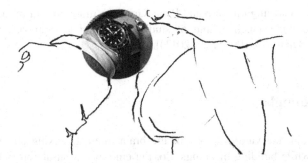

Fig. 5.15 The Submariner Rolex on the wrist of Sean Connery in *From Russia With Love*; the same product was placed in *Dr. No* and in *Casino Royal* with Daniel Craig (drawing by the author)

Fig. 5.16 The Aston Martin DB5 appearing in *Goldfinger* and in *Skyfall*, with the same plate

Fig. 5.17 Heineken beer in the hands of Daniel Craig in *Casino Royal*; Smirnoff Vodka has been placed in many movies: after *Quantum of Solace*, with Daniel Craig, a special edition has been produced, and with the claim "shake, not stirred" as always asked by James Bond; other similar products were placed, as Martini in *Diamond are Forever* and Heineken in *Skyfall* (drawing by the author)

that metaphors carrying on Care and Panic/Grief are more present in the sedentary agricultural cultures [such a kind of study could start by approaching (Cavalli Sforza 1981; Diamond 1999; Sica 2011)].

5.3 An Example

As already said, the various levels of the brain interact, mixing all the kinds of perceptions while building meanings. So, the emotions arousal comes from many different signals, coping with primary perception, basic and complex metaphors; in order to explore the relationships between simple perception, metaphors and emotions, we carried on the following twofold experiment.[12]

5.3.1 Analysis of Handles

We presented to the students the following set of handles, asking them to qualify each product, according to a predefined grid (Fig. 5.18).

The qualification grid asked to position each handle according the following exclusive attributes (the posed question was "In whose house do you think will find the handle? Which kind of final user will choose it?")

- *gender*: male or female;
- *age*: young (till 40–45 old) or elder (definitely more than 55–60);

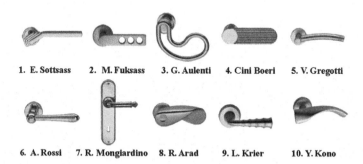

1. E. Sottsass   2. M. Fuksass   3. G. Aulenti   4. Cini Boeri   5. V. Gregotti

6. A. Rossi   7. R. Mongiardino   8. R. Arad   9. L. Krier   10. Y. Kono

Fig. 5.18 A set of handles by well known designers: the handles no. *1, 3, 4, 5, 6* and *7* are in *yellow* brass

---

[12] The experiment has been carried in many different workshops, in Italy and in Mexico, with an international audience, mainly of graduated students; the results we refer have been collected from a hundred of persons in total.

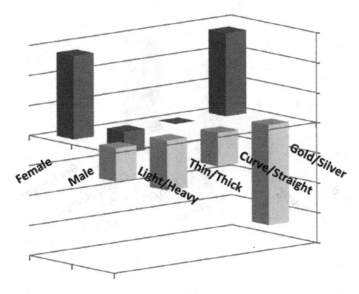

Fig. 5.19 Perceptual properties against gender. The *gold* coloured handles are suitable for women, while *silver* is for men; *light* structures for women and *heavy* for men; *straight* lines are more suitable for men than for women, but not so strictly. Warm colours and *light* structures recall soft sensations, we can connect to Care; *cold colours*, heaviness and *thickness* recall aggressive sensations, we can possibly connect to rage. Behind the selections two usual stereotypes seem appear

- *life style*: oriented to modern times or to tradition;
- *social level*: upper, middle or lower class.

The results of the experiment shown that there are specific perceptual characteristics strongly related to a stereotypical view of the potential user: the colour, the presence of curves, the thickness of the structure and the shape suggesting heaviness.

The found relationships are described in the Figs. 5.19, 5.20, 5.21 and 5.22.[13]

A further question asked to the participants was: which profession do you think for such a handle? The answers were clarifying the stereotypes and the metaphors: a lawyer was related to gold and heavy, i.e.

---

[13] The observations have been carried on in different countries, with different persons coming from different nations, and in different conditions. We cannot consider this experiment methodologically satisfactory from the scientific point of view; nevertheless some useful remarks can be done. The scales are purely indicative; a high bar in the positive direction indicates a strong evidence that an attribute (e.g. golden colour) has been related to a qualification (e.g. female); the opposite of the same attribute (e.g. silver colour) is represented as a negative value.

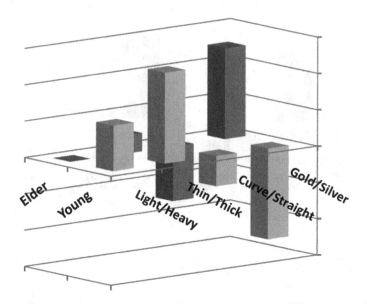

Fig. 5.20 Perceptual properties against age. The *gold* coloured handles are suitable for elderly, while *silver* is for younger, as well as for *light* and *thin shapes*. Again, stereotypes appear: younger are more essential, decided, innovative

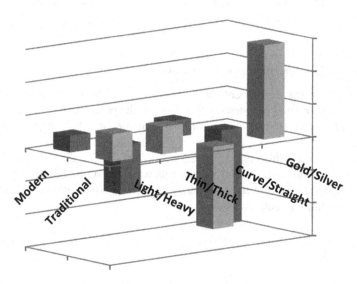

Fig. 5.21 Perceptual properties against life style. Traditional is *thick* and *gold*; *modern straight*

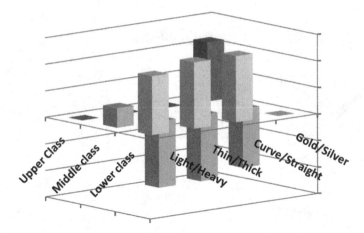

Fig. 5.22 Perceptual properties against social status. *Upper class* is gold; *middle class* is heavy, thick, straight; *lower class* is curve, thin and light. The quantity is stereotypically the goal of a *middle class* climbing the society

Fig. 5.23 Mike Jagger, leader of the rolling stones, and Pamela Anderson "sexy" actress: which best handles for them?

- an architect, an engineer, a professional is male, young and middle class, according to silver, straight, thick;
- an artist is typically a female, elderly, upper class, according to gold, curves and heaviness

and so on.

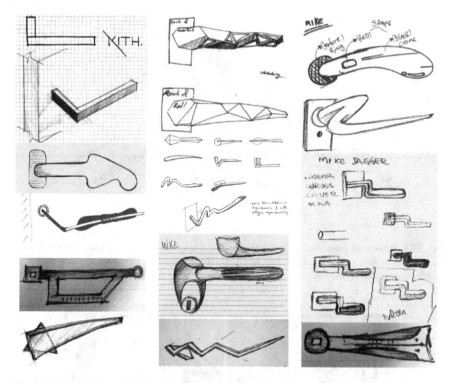

Fig. 5.24 Handles sketched for Mike Jagger

It seems that the answers are not related to the single perceptual signals, but to more complex clusters: Whose belongs the handle no. 3? May be an artist, female, successful (and then not young), then belonging to the upper class: the stereotype provides metaphorically the attributes.

This hypothesis seems to be confirmed by the experiments on the synthesis.

5.3.2 Synthesis of a Handle

We presented to the same classrooms a couple of well known public persons: Mike Jagger and Pamela Anderson, and the students were required to design the proper handles for their homes.

All the students, without any exception, interpreted the two characters in the same way; the former: male, strong, decided, sharp, successful, confident in itself (Rage); the latter: woman, soft, exciting (Care and Lust). The Jagger's handles were sharp, with edges, essential, sometimes with references to guitars or microphones; the Anderson's handles, curve, soft, with decorations, sometimes blinking to her

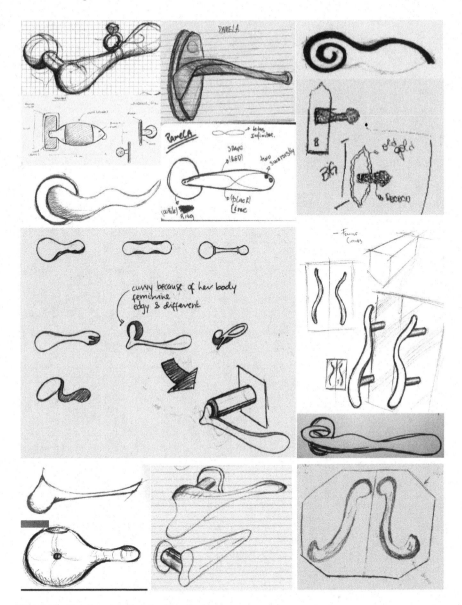

Fig. 5.25 Handles sketched for Pamela Anderson

body or luxury research. The character defined the emotions to be expressed, and then specific shapes were chosen (Figs. 5.23, 5.24 and 5.25).

Chapter 6
The Design Process

Abstract The chapter discusses some typical design process as declared by Masters, and proposes possible integrations for taking into account perception and emotion. After the evaluation of the power of the emotions in modifying buyer's behaviours, the problem of the designer responsibility and ethic are analysed. Examples are provided.

Keywords Design process · Emotional design process · Responsibility · Ethics

6.1 Is Design Moving Toward a Science?

We are able to teach someone to paint, to manage oil colours, to use perspective rules, and so on, but we are not able to teach someone to do a master piece. It depends on the fact that we have a strong knowledge of painting materials, techniques and methods, and we largely ignore what art is. The Renaissance ateliers were places in which a master was able to teach a pupil about techniques and methods, to improve his/her knowledge about the current philosophy, and so on; but he/she was not able to teach to produce masterpieces, despite the pupil was able to learn it, from examples and through imitation of the master. But the technical and methodological aspects were well found, systematically organized and described.

Design is in a similar situation: design schools teach a lot of tools and techniques, and provide a quantity of general cultural knowledge, but are unable to create the master. Moreover, according to the common feeling, the designer is seen mainly as a creative person, doing "strange" and interesting things.[1]

---

[1] In a recent experience, about one hundred and fifty students attending the second year of a Communication Degree Master were asked to provide three posters emphasizing the three souls of the Politecnico di Milano: engineering, architecture and design. More than the 75 % of the posters aiming at representing the design course values pointed out creativity (expressed either directly with the word "creativity" or in terms of colours, improvisation, exterior appearance, and so on; moreover, more that the 10 % propose as a goal the individual success); by the way, not only "creativity" is not taught at all at the Politecnico, but, according to the author, it is not at all possible to teach it.

© The Author(s) 2015
M. Maiocchi, *The Neuroscientific Basis of Successful Design*,
PoliMI SpringerBriefs, DOI 10.1007/978-3-319-02801-9_6

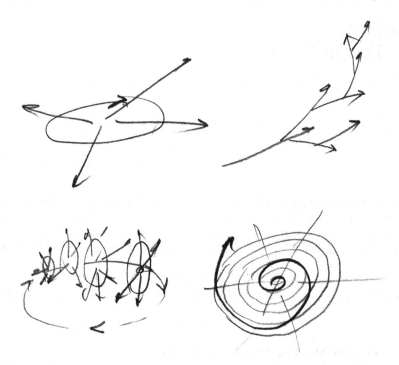

Fig. 6.1 Some sketches on the design process by Enzo Mari [reworked by the author from Mari (2001)]

Nevertheless it is mandatory for a school to teach not only techniques, but also a solid method.

Design literature presents many different approaches, and we consider the two proposed by Mari (2001) and Munari (1981) particularly interesting (Figs. 6.1 and 6.2).

As suggested by Oh (2013), we can consider the former as the *poet of the project*, and the latter as the *engineer of the fantasy*. In front of a need, Mari explores various alternative roads, often changing his direction, until a satisfactory way seems to appear; this way is then deeply explored, studying the details at increasing levels; restarting sometimes the process with some backtracking. It seems an approach based on a huge capability and experience, but mainly related to creativity, as ants, going here and there, can find some interesting elements on their way. Mari is not progressing by chance, of course, but it seems that experience, analogy, lateral thinking and a deep mastering of the matter allow him to get excellence. This approach can be learned by pupils, but cannot be taught by a teacher.

Munari's approach is more similar to the one of the engineers: as depicted in Fig. 6.2, the process involves many precise steps, one after another, and creativity seems to spring from the constraints imposed by the context: what is not forbidden is possible. After the problem definition, he proposes the identification of all the relevant elements and the needed functions, then the analysis of any kind of constraint: the physical and psychological ones allow to identify the fundamental

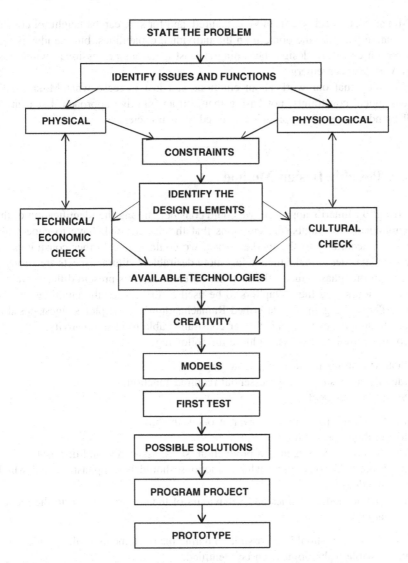

Fig. 6.2 The design process proposed by Munari [modified from Munari (1981)]

design elements, while the technical, economic and cultural ones could cope with the technologies to be used. At this point, all the defined constraints can free creativity: what is not forbidden is allowed, and from them some models can be suggested. A first mental verification can be applied, in order to compare the various possible solutions (among which the simplest ones should be chosen). Finally, the project and a first prototype. Of course, the process is not linear, but at any step some observations can suggest changes or backtracking in the already done activities.

Munari's approach seems more structured, and for sure can be taught; of course, the creative parts are the domain on the individual capabilities, but the idea is that any person can learn design, becoming at least a serious professional, while possibly very few can emerge.

We think that our work about emotions can just increase what Munari calls psychological constraints, not basing them on an intuitive approach, but using a well founded scientific approach suggested by neurosciences.

6.2 A Possible Design Method

Referring to Munari's approach, we could consider, among the identification of the psychological constraints, the emotions that the artefact will have to cope with; moreover, according to the market target, we could define further constraints on emotions, in order to make the artefact more desirable. Cultural aspects are strongly relevant in this phase, due to the fact that different cultures present different values and behaviours, and the metaphors to be used as drivers for the emotions can be quite different.[2] Again, the described Ramachandran's principles suggest possible shapes to support or to avoid: further constraints able to raise creativity.

So, the complete process could be the following:

1. State the problem and identify the target;
2. Identify the issues and the essential required functions;
3. State the constraints:

   (a) Physical (size, weight, power, eco-sustainability, etc.)
   (b) Technical feasibility;
   (c) Economic constraints, in production, sales, assistance and disposal;
   (d) Emotional constraints: which emotions should be emphasized and which avoided;
   (e) Characteristics, values and behaviours of the cultures related to the market target;

4. At this point, it should be possible to define many elements of the project, and, the possible technologies can be identified.
5. Now, creativity can be helped by the studies of emotions, to better define the aspects of the final artefact:

---

[2] A deep analysis of the different values and behaviours in different cultures is carried on in Hofstede et al. (2010); the study shows how different communities (we can approximate as different countries) accept or do not accept some social phenomena, such as specific social roles of males and females, or excess of different sustainability of the social classes, and so on; for example, a smart phone covered of Swarowski cristals, adding nothing to the functions, but increasing enormously the price, can be accepted in China and is refused in Danemark: China is accepting the social differences more than Danemark does.

(a) It is possible to do it by targeting directly the reptilian brain or the limbic system: edged lines, aggressive faces, rounded shapes, soft tones, and so on can work, but, as we already observed, those aspects seems related to Seeking only, and we need more;

(b) It is necessary to start from the upper levels, accurately selecting metaphors we can relate to scripts, able to stimulate the pleasant emotions (care, lust, play, rage, seeking), and providing in some way the perception and the communication of such metaphors; the choice of the metaphors will be connected to the cultural aspects;

Of course, these aspects are only the final solution, and the personal experience and sensitivity, and lateral thinking play a relevant role in finding the best solution; nevertheless, in this way we are able to define more constraints, and then to foster creativity;

6. Then the following phases will proceed as in Munari's model, providing that the tests on the model can be carried on also taking under control the compliance to the emotional constraints.

6.2.1 An Example

As a short and simple exemplification of the method, we present the following simple simulation.

Brief. Design a new kind of pasta, easily recognisable and different from other kinds, so than it can be considered a unique unusual product, and not a choice by comparison with other similar goods.

According to the brief, it appears that the main goal is the need to avoid comparisons. So we should not produce a better kind of pasta, but a new one. Of course, the quality of our product is relevant, and it will not be lower than the ones presently available on the market.

For the above reason, not only the quality of the raw materials will be high, but also other required technical characteristics should be taken into account, such as the capability to catch sauces, the ease of use, the size.

Following those remarks, we can add some constraints to our project:

- Pasta for everybody, adults and children, and for any country.
- According to the fact that this pasta must be clearly recognizable as different, we can orient to an upper class consumer.

Now we can start with the various steps:

1. Emotion and values. The typical consumer could be represented through the stereotype of an upper class family, with good spending capabilities, good cultural level, needs to feel themselves different, able to understand the qualities; the mother, manager, dresses branded classic wears; the father is an important

manager; both are about 40–45 old; they have two cars (one of the upper segment, the other smaller but of the same luxury level); they live in a large apartment, provided with any kind of comfort; they have one child, male, 9 years, with good results at school, practicing some sports, well educated, with few friends. Of course, it is not the typical consumer family, but it is the typical appearance for our target. Then, what the product should cry is: if you buy me you are *socially upper, serious, culturally better, more sophisticated than the average, careful to quality, concerned about health.*

2. According to the above definition, the following attributes of the shape could be derived:

   (a) Soft lines without edges;
   (b) Surfaces suggesting soft textures;
   (c) Non-aggressive lines;
   (d) Recall to tradition;
   (e) Recall to history;
   (f) Pasta as Italian product with Italian design;
   (g) Warm colour (possibly more dark than light, without reaching the yellow of the eggs-pasta, but not too white; no vegetable colours at all, like carrots, spinaches or other);
   (h) Possibly, a family of pastas, with different sizes, but described as suitable for different uses (sauce consistency);
   (i) Structure of each piece easy to grasp for children and adults (also with the use of chopsticks);
   (j) High quality of the raw materials, and no difference in the thickness in the various parts of each piece, in order to keep the cooking "al dente". So we excluded kinds such as *spaghetti* (non easy to eat) or *farfalle* bended in the middle, then with problems in uniform cooking.

3. Now we have to cope with Ramachandran's principles. We should avoid peak shift (it could result in some kind of caricature, in contrast with the adjective serious); perceptual grouping could reduce the identity of the single piece; abhorrence of coincidence/generic viewpoint could be not sufficiently distinct; the same for contrast (aggressive); we could use isolation, perceptual problem solving, rhythm, symmetry, orderliness, balance.

4. Then we have to cope with the metaphoric structures; according to the above analysis, we should emphasize cultural aspect, tradition, history, Italy. So we can start to put on a sheet some keywords: *design, Italy, tradition, culture, history*; being Italian design more recent and not with tradition on pasta, we start for the moment to discard it, trying to connect in a network the other elements; the first new keyword coming is *Renaissance*, and, as more representative, *Leonardo da Vinci*. Then we try to find some relationships between Leonardo and shapes. A quick research on Google presents the drawings by Leonardo da Vinci for the book *De Divina Proportione* by Luca Pacioli. We connect every element in a network, as we could use it in order to communicate the new kind of pasta, that, of course, will have the shapes of such solids (Fig. 6.3).

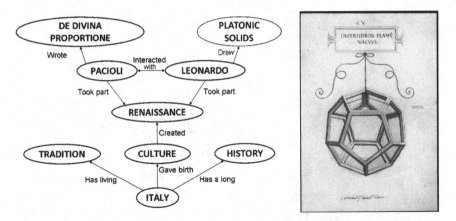

Fig. 6.3 A semantic network on the relationships described in the text. The idea is to use the net in the communication (e.g. on the packaging), in order to make the user feel him/herself as a person able to understand and to appreciate so fine and deep concepts, at to take part in a cultural process. On the *left*, an example of Leonardo's drawings

So, the new pasta is defined.

- A collection named: *De Divina Proportione*, with four different formats: the cubes, the octahedrons, the dodecahedrons, the icosahedra; tetrahedrons have been excluded for hard edges present;
- their size will be different, according to different uses, but, being structured as thin cylinders replacing just the edges (as in the drawings by Leonardo), the cooking time will be the same;
- different sizes and internal space can orient the best pasta for the best sauce (tomato for the smaller, Genoa pesto for the middle-size, meat ragout for the bigger), but not in an exclusive way;
- all of them, mixable, are easy to be caught with a fork, or chopsticks, also for children;
- the colour will be light brown-yellow; the edges will be slightly rounded, the surface will be slightly knurled;
- symmetry, rhythm, orderliness, balance are deeply embedded;
- the packaging will recall the classical world in fonts, the drawings by Leonardo will appear on the box, some news about Leonardo, Pacioli and the Renaissance will be present, together with the comparison between the harmony of the Divina Proportione and the harmony of the taste, the shape, the best quality of the ingredients, the equilibrium in nutritional facts, etc.

Figure 6.4 presents a sketch of the result.

Note that, to avoid too many problems in moulding pasta, some of the edges are cut, in order to avoid complex and expensive machines and processes.

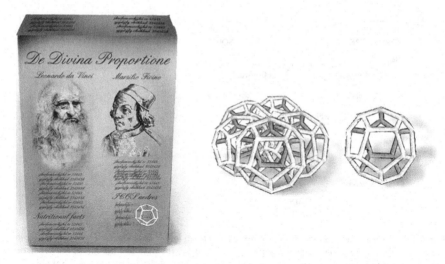

Fig. 6.4 Example of pasta and packaging

6.2.2 Some Warnings

The presented example is not the demonstration of some kind of automatic design production: all the relevant choices are always in the hands of the designer; simply, we believe that a better knowledge of the power of the various perceptual communication mechanisms, as well as the way in which they work, helps to increase the consciousness of the design choices. We are not suggesting that design can be algorithmically developed, but we hope, modestly, to move a little bit far the boundary of what we know and what we do not in this field, always bridging technology and art, engineering and creativity, needs and dreams.

Nevertheless, the power of emotions and the related autonomous unconscious arousal can be a dangerous element in persuading the customers to buy products; quoting Papanek (1971):

> Advertising design, in persuading people to buy things they don't need, with money they don't have, in order to impress others who don't care, is probably the phoniest field in existence today.

The more the designer understands on the user brain, the more he can influence behaviours. For this reason, the responsibility of the designer is becoming more and more relevant.

Fig. 6.5 An F1 Ferrari with a barcode recalling Marlboro, a hidden message based also on the colours red and white; Ferrari accepted to remove the barcode. The same problem occurred in MotoGP' Ducati motorbikes, always with the Marlboro barcode (detail from a *photo* by Jeff Wunrow, CC BY 2.0)

6.3 The Evidence of Responsibility

The more the designer increases his/her knowledge on the design process and on the characteristics of the final result, the more he/she increases his/her power towards the buying pressure on the final user.

This is widely shown in the works of Lindstrom (2010): through brain imaging techniques, he shows how it is possible to influence the consumers through the products appearance or through communication; for instance he shows how the red colour of a Ferrari in a F1 race, together with some white spots imitating a bar code (Fig. 6.5), can bypass the bans on the cigarettes ads; how Pepsi Cola brand activates the "flavour" brain centres, while the ones of Coca Cola is related to the membership in a group; how a proper choice of a picture on the tobacco packaging, showing terrific images on the effects of cancer, can be a good selling point, instead of an obstacle, and so on.

So, it is evident how ethical rules on avoiding subliminal messages[3] are easily overtaken by other "overliminal" techniques: metaphors and Ramachandran's principles are just some examples.

The ethic responsibility of a designer is so very high, being he/she, voluntary or not, influencing non rational behaviours of the consumers. The responsibility is not simply related to the fact that the consumer's freedom in buying is influenced, but in the fact that other messages are carried on, influencing cultural values; for example, many advertising messages support a gregarious role of women as submitted to men and considered as objects (Fig. 6.6).

---

[3] A subliminal message is a signal with characteristics that are lower than the perception thresholds; for instance, an image lasting less than 1/20 of a second, a sound lasting so short time, to be impossible to get the proper air wave pressure, and so on. In proper terms, product placement techniques (e.g. showing cigarettes or alcoholic in the movies) are not subliminal, beside the fact that the spectator's attention is distracted by other events.

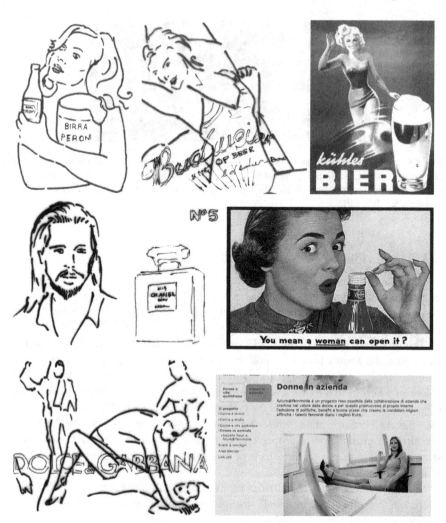

Fig. 6.6 Ethic values in communication (*top to bottom, left to right*). First row, beer and women: the woman is pretty, blond, to drink, to be consumed; the women according to chanel perfumes, "dressing" Chanel to catch the best men (and this is what a woman should tend to); Tomato sauce: the woman is weak, and usually cannot survive without men; Dolce and Gabbana: a man raping a woman, with an attending group; *futuro@femminile* (future at feminine), a website supported by the Italian Government to push women to study informatics (very few females attend to those courses): the provided model is the worst arrogant masculine manager (drawings by the author)

Maybe those messages work from the point of view of sales, but for sure they reinforce stereotypes, unacceptable at least for the actual western countries.

Responsibility means to exercise our freedom, evaluating, considering positive aspects and accepting the consequences of our actions. So while speaking about responsibility we have to deal also with action and freedom. We can represent the situation with the diagram of Fig. 6.7.

Fig. 6.7 Responsibility, freedom and action

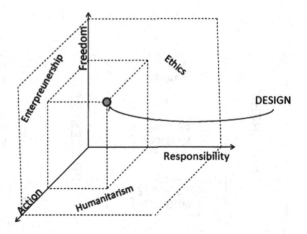

Most of the actual cultures consider personal freedom as a vital value: we all want to be free to act, free from constraints, but freedom cannot be absolute, being limited by the other's freedom. Moreover, freedom can be just a philosophical idea, but embodied through actions.

Action in itself is a value, able to change the status of the things surrounding us: without actions nothing happens.

In this complex space, moving on the surface *freedom-responsibility* without action means to think about Ethics as a Philosophy branch. On this space it is possible to discuss about what is good and what is bad, which limits are proper in actions, but without being the engine for changes.

While moving on the surface *freedom-action*, the risk is piracy: we want to be free to do anything, without responsibility towards others; the best possible interpretation is a good entrepreneurship, but lasses-faire economic theories seems to fail in respect to an harmonic growth of any nation and any classes.

When we move on the surface *action-responsibility*, without freedom, we are ruled from humanitarian constraints.

The designer requires the will to act, in order to create, products, services, material or immaterial artefacts, requires the freedom of any possible solution driven by his/her creativity, must have the responsibility to take into account the consequences of his/her projects, in terms of any possible stakeholder, including the whole community, and the environment, including the near and far countries, including the present and the future. This is the space of design: the space of freedom-action-responsibility.

The word *design*, spreading in the last years over any application field, seems so to recover the original meaning of a project in general; this brings back memories of virtuous entrepreneurship, as proclaimed by the european foundation for quality management (EFQM) (Fig. 6.8).

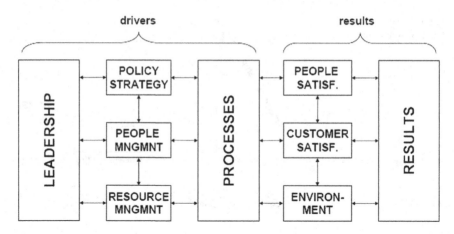

Fig. 6.8 The European Foundation for quality model of a virtuous company

According to the model, the goal of a company must be the economic result: losses will be paid unavoidably by someone, and will demonstrate that it is a 'dissipating' machine, destroying values instead of creating them.

The economic results are strongly related to three elements: the satisfaction of the personnel (unsatisfied employees will work without adhering to the company goals, will leave the company, and the best elements first), of the customers (no customers, no company), and of the social/ecologic environment (this is a way to measure the balance sheets both geographically and historically: for example, pollution will cost for future recovery more than the actual margins, i.e. to pollute means to steal from future; to exploit poor countries through low cost workers and through the control of their internal market corresponds to transfer richness geographically, without value creation, and so on).

In order to get those results, the model indicates a strong leadership, able to communicate and make transparent and shareable policies and strategies, able to define the proper respectful and favourable management of the employees, the proper and shrewd management of the resources and of the commodities; more, according to the model, the key point for the proper quality of the final result is the definition, knowledge, measure and evolution of the internal processes.

This all is design with responsibility.

Chapter 7
Case Studies

Abstract The chapter presents various examples of projects, some of them really implemented, some just as projects, some as products of teaching laboratories, all of them coping with the design of the emotions for the users. The projects have been carried on mainly in health care environments and for public transportation systems.

Keywords Emotions design · Emotional design practices · Emotional design examples · Emotional design for health care · Emotional design for public transportation systems

The models described in this text are the result of many researches over the last years: real experiences induced thoughts about what design is, local project needs suggested to investigate perceptual aspects, studies on synaesthesia influenced the research directions, and so on.

Up to now we had no chance to build a project taking into account all the aspects of the model, but the activities carried on, on the basis of some of the related aspects, deeply influenced its development. In the following pages we provide some examples of the works done, none based on the complete final presented view of emotional design, each of them involving a piece of the presented knowledge.

7.1 A First Experience in a Children Hospital

In the Children Cancer Hospital "Pausilipon" (Naples), the responsible for the Radiology Department of the Hospital, prof. Enzo Salvi, decided to change the exterior look of the Nuclear Magnetic Resonance machine. This equipment has usually a threatening shape, imposing moreover to remain still, while its use produces an annoying noise (Fig. 7.1). For the above reasons, it is quite difficult to keep the children quiet, and, in order to avoid too many ineffective results, sedatives are given to children in about the 40 % of the cases.

Prof. Salvi asked the artist Mimmo Paladino to do something in order to change the look of the machine, transforming it into some playful object. He decided to

© The Author(s) 2015   75
M. Maiocchi, *The Neuroscientific Basis of Successful Design*,
PoliMI SpringerBriefs, DOI 10.1007/978-3-319-02801-9_7

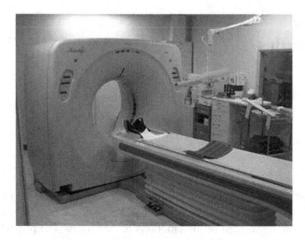

Fig. 7.1 A typical nuclear magnetic resonance equipment

Fig. 7.2 Prof. Salvi and the decorated nuclear magnetic resonance equipment, with a detail

cover the surface of the machine with many repetitive, rhythmic shapes, without affecting any of the commands of the equipment (Fig. 7.2).

The emotional impact of the machine changed totally; according with the principles of Ramachandran, the rhythms and the repetition of the figures stimulated the children positively, and their Seeking distracted them from the examination, making them more quiet and still. The result of this simple intervention is the fall of the sedation cases from 40 to about 2 %. The collateral effects of the emotional change are then, not only a strong reduction of the invasiveness of the sedatives on the patients, but also a significant reduction of the operational costs (less sedation costs, less number of examination to be repeated, and then less time spent by the staff), in front of a very cheap intervention (in this case, moreover, Paladino offered his work free of charge).

Enzo Salvi decided the intervention on the basis of his experience and sensitivity, but in fact applied principles of Emotional Design.

This first experience suggested to investigate the chances to work in health care environments, in order to affect the behaviours of the patients through changes in the physical environment, with actions able to increase positive emotions and to reduce negative ones.

7.2 Emotional Design in Health Care Environments

Between 2007 and 2011 a strong co-operation was established with the Istituto Nazionale dei Tumori di Milano, a research and care centre on Cancer. The goal of the works was to experiment changes in the non-medical aspects of the environment, in order to improve the experience of the patients, with the strong believe that such a kind of actions would have made the therapies more effective.

The approach followed defined the various steps of a "journey" in the care road:

1. Registration
2. Visit
3. Exams
4. Diagnosis
5. Check in
6. Treatment
7. Life
8. Visit
9. Check out

For each step, we examined the different characteristics of the physical environments, the actions carried on by the patients, and their interactions with the staff and the structure. Then we defined which kind of actions was possible to apply either in the physical environment (furniture, walls, colours, etc.) or organisational (kind of interactions, interfaces, wayfinding, etc.).

All the proposals were oriented mainly to increase Seeking, Care and Play.

Unfortunately, the absence of funding and the impossibility to apply a systematic approach prevented us from getting measures of the effectiveness of the actions, and the only (weak) measure we had, for any of the actions we are going to describe below, were the growth of the smiling people (both among the patients and the staff), a collection of written positive comments, and some interviews.

7.2.1 Before Accessing the Istituto dei Tumori

The website of the Istituto provides much information about access, visits, examinations, booking and so on. Moreover, statistical information has been added on
the typical waiting time for an appointment or for a visit; the user can verify how
many days or how much time it will be possibly necessary to wait once he/she
arrives in the proper place.

One of the bad situations in which the patient is emotionally involved is to wait,
without any information about how long. The reduction of the uncertainty is a way
to reduce Grief, and then to increase Care.[1]

7.2.2 Information and Wayfinding

Once a patient enters the large reception hall, many services are available, and not
always in a well recognizable way: information points, the Unit of Customer
Relationships, various voluntary associations, and so on.

We set up a computerized info-point, in a central position, for which we studied
the user interface (icons, colours, shapes, language level and rhetoric) (Fig. 7.3).

Fig. 7.3 The info-point, with the initial screen

---

[1] Recently we observed interesting solutions in such direction at the health care centre Humanitas,
in Rozzano (near to Milan): the people accepted by the First Aid, provided with a numbered ticket,
can see on a screen the estimated waiting time for the service.

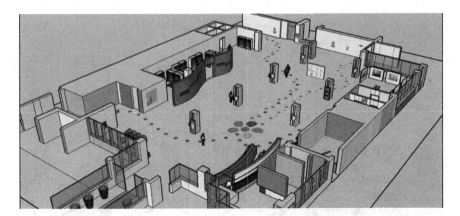

Fig. 7.4 The layout of the main hall, with the coloured paths

We selected the icon of a puppet, with a round and big head and always smiling (Peak Shift), as a guide for the queries, in order to suggest a relationship with a person, soft and happy, in order to increase Play and Care.

A further function was prototyped but not set at work: the possibility to download on the patient mobile phone (through a bluetooth connection) the map of the environment, with the path to be followed to reach a specific place. This function too had to increase the experience of Play and Care.

The same background on the screen was supposed to be drawn on the floor: a flower with differently coloured petals, indicating the paths to be followed for reaching the required services (Fig. 7.4).

7.2.3 Waiting Rooms, Visits, Examinations, Therapies in the Breast Radiology Department

We had the possibility to reorganize a whole area, the Breast Radiology Department, during its refurbishing.[2] We had many constraints: the technical offices already defined the contractual aspects of the materials and of the furniture to be employed, and the unique freedom degree were related to colours and positioning.

Nevertheless, the environment obtained is quite different from the ones you can see everywhere in a hospital, and every visitor recognizes the astonishing difference immediately.

---

[2] As for most of the projects here described, they were designed and implemented by the students of the Courses of Design of Communication of the Politecninico di Milano. The names of the authors and an extended description of the projects can be found in Maiocchi (2007, 2008, 2010a, b).

Fig. 7.5 The waiting room, with the paintings

A further chance was given by a request to some painters to provide as a gift a painting properly devoted to the specific environment. The few friends that accepted the invitation mailed a similar request to other colleagues, so that within a couple of months we received 170 masterpieces from 160 artists over 7 countries.

The paintings became a central element of the refurbishment.

The walls were painted with white nuances, in order to enhance the contrast (see Ramachandran's principles) with the coloured paintings, but the waiting room, mainly in orange. The doors were chosen on the basis of a colour code according to the role of the rooms (studies, dressing rooms, examination rooms, etc.), and the colours were strong and unusual for a hospital (acid green, deep purple, red, and so on); the seats of the waiting room were purple, and their position not completely ordered, symmetric but not boring (Figs. 7.5 and 7.6).

All the choices were oriented to enhance the arousal of Play and Seeking. Moreover, in order to collect information and to induce some Care, carnets and pencils were available to leave comments, and some bookshelves allowed book-crossing activities and reading.

7.2.4 Check Out

After the check out, a cancer patient is not properly healthy: he/she must be periodically controlled, and the regularity of this occurrence is dramatically relevant for his/her future. Unfortunately, too often the patients prefer to avoid the subsequent needed controls.

Fig. 7.6 A corridor, with paintings

Fig. 7.7 The "check out" passport

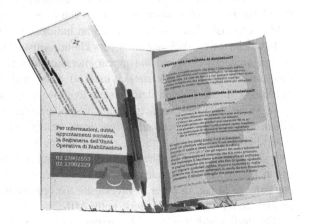

We introduced a kind of "passport", testifying the check out, and, in addition, establishing a psychological connection between the patient and the centre (Fig. 7.7).

In this way, the relationship between patients and structure is inspired by a Care emotion, able to induce them to keep the contact alive.[3]

---

[3] We discussed many times on how to measure the effectiveness of the undertaken actions; beside the best practice of considering the environment as a part of the health care environment, and then of applying well known methods, as it is done for the drugs tests (statistical methods based on double-blind approaches, with placebo subministrations), the doctors of the Istituto dei Tumori suggested to record the compliance of the patients behaviours to the raccomendations of their doctors. In fact, one of the recognised problems is the fact that the patients disregard what it is

7.3 Observing and Modelling Human Behaviour to Design Services in the Transportation Systems

7.3.1 The Goal and the Constraints

Underground stations are core points for urban social life. Every day, a large number of citizens cross the access areas leading to trains, spending a non negligible time in transport activities; even in a middle size city like Milan, some intersection stations count daily more than 100,000 travellers and the efficient working of the service influences significantly the quality of the life of individuals and of the community as a whole.

While travelling, users live an experience that is produced by several different components:

- the efficiency of the transport process;
- the quality of stations and train physical environments;
- the interaction with the company employees and other travellers;
- the presence of subsidiary information and services, and so on.

Furthermore, the quality of the travel experience plays a non-negligible role in the global image of a town; it can be taken as a measure of the efficiency of the public administration government and an indicator of the city values.

In several towns, such as Zürich, Stockholm, Amsterdam, the underground stations are magnificently designed and appear interesting places worth to be visited; in other cases, such as in New York, stations have often essential dimensions, and consist only of a simple thresholds and stairs system to allow access to trains (Fig. 7.8).

In Milan, like in several other towns, underground stations have usually large dimensions but are conceived as functional spaces, and in most cases offer quite a modest appearance although clean and sufficiently comfortable.

In 2010, our institution, the School of Design at Politecnico di Milano, together with Milan School of Art, Accademia di Brera, cooperated with ATM, the local transport company in Milan, in order to investigate the opportunities of improving the travel experience for underground users, in the MIND—Milan Network Design project.

The design activities faced a number of tight constraints:

- *Safety issues* Due to topological characteristics and symbolic factors, the whole underground system is considered critical from the safety point of view. The different aspects of security include severe factors as fire events, terrorist attacks, crime phenomena, and so on.
- *Non-interference* with the main performances of the transport service. As the core performance is transport, individual activities should not be interrupted or

---

(Footnote 3 continued)

asked to them. So, this "passport" is not simply an emotional gadget, but an element able to change the behaviour of the patients, making them "more" in touch with the centre, as needed.

Fig. 7.8 An underground station in Stockholm and in New York

delayed and designed solutions should not interfere with the normal traveller flows. Individual needs of privacy should also be taken into account in the design of interactive solutions.

- *Non-invasive character* Designed solutions should not be invasive or require significant physical changes of interiors. This constraint rises from economic reasons and other factors related also to way-finding issues and to the relevant role played by advertising postings.
- *Acceptability requirements* As the public transport service is a public facility offered all citizens, it is important to take into account acceptability needs in the widest sense.

7.3.2 Observation

Given the constraints, it seemed mandatory to organize an activity of intense analysis of all the factors influencing the users' experience, also involving the students participating to the project courses. The design final goal was the creation of solutions capable to modify the emotions related to travel activities, without interfering with them and without modifying the physical characteristics of the spaces; in order to produce events, interstitial, but relevant from the emotional point of view, we first had to better understand the complex of emotions that are relevantly associated to the underground travel experience. Then, the solutions had to be designed in terms of services, i.e. nonmaterial interactive systems offering functionalities, information and supporting user activities; the main goal was not the satisfaction of practical need, but the improvement of users' attitude and feelings towards the local context and the transport system as a whole.

To produce effective solutions, we needed a better understanding of emotions and feelings related to the fruition of the underground service and to the physical environments (station spaces, platforms, train interiors, and so on).

The analysis phase was carried on with a variety of techniques:

- ethnographic research in the field in the tradition of contextual design (Beyer 1998) including interviews, forms, activity observation and modelling;
- self-reflective analysis of personal experiences as traveller;
- quantitative analysis of social behaviours also including the tracking of most practiced routes, interaction between strangers, space occupation, walk speed in the different contexts an so on;
- observation of measurable factors usable as indicators of people feelings and emotions: postures, gestures, facial expressions, walk speed in the different areas of stations, availability to conversation and so on;
- analysis of people reactions in presence of purposely organized performances;
- collection of extreme cases of observable events and behaviours taking place in Milan underground stations and trains;
- collection of digital documents available in social networks (YouTube, Twitter, …) providing hints of events, opinions, personal stories related to the underground services around the world;
- collection of video excerpts extracted by popular movies and set in an underground environment.

As the observation in the field played a fundamental role in the understanding of the underground context and phenomena, we organized them carefully. The service opening hours were divided into time segments characterized by homogeneous crowding conditions. The variable observed included face expressions (for which a reference palette was previously defined), body attitude, activities, walk speed.

The great amount of information gathered during these activities was analyzed in order to ascertain the most common mental attitudes of travellers during the trip experience. A number of behaviour trends, personal strategies, recurring situations and social positions were brought to evidence; the population travelling in the underground was classified in a number of different *ethnic groups*, each of them characterized by a mind attitude, a physical behaviour, or a strategy.

The classification confirmed a quite diffused tendency to self-insulation for most lonely travellers: as expected, depending on the time of the day, boredom, anxiety, sensitiveness were quite diffused, since life style in Milan is quite frenetic; furthermore, observation and experiments revealed a scarce attention toward commercial advertising and other visual events animating the environments.

On the other hand, a number of personal *survival strategies* were evidenced providing interesting inspiration to project. Our underground seems to be populated by *readers, space invaders, hygiene fanatics, sleepers, chat catchers, workers, music listeners, phone users*, and other species capable to invent practical solutions to feed their personal needs during the trip. The trip experience seems to be much more pleasurable for people that have the opportunity of social groups; small or big groups use travel time to talk about a variety of subjects: work matters, personal confidences, sports, shopping wishes and so on. Also casual conversations with unknown partners (when not roused by unintentional offense), seem to bring positive effects on the atmosphere, also influencing the neighbour area attitude. Quite controversial is the effect produced by animating events, such as the presence of music players, beggars, noisy people or lively children.

For instance, as confirmed by the observation data, the walk speed of several passengers (mainly women) accelerate when entering the descending corridors; several women change the bag position entering the underground; the space disposition of people and personal postures change when the train leaves the underworld entering the open-air parts of the route.

While it is impossible to identify and measure absolute mind disposition, physical changes can be taken as a signal of a modification of emotions, attention, and attitude especially when they show systematically. The modelling of people behaviour and of emotional attitude was supported by knowledge provided by brain sciences and neuroscience literature explaining human conscious and non-conscious mental processes, empathy phenomena through mirror neurons effects, decision processes and emotional mechanisms related to actions.

In most cases, the behaviour of individuals seems to be quite affected by the presence of other travellers, and most people seem to be willing of being considered coherent with the dominant not noticeable attitude. In other words: people tend to avoid any kind of contact with other unknown passengers and try to develop strategy to avoid conflict or interaction.

Observations also demonstrated a diffused sense of anxiety generally related to the permanence in the underground contexts, probably related to the specific topological conditions of confinement in closed tunnels. This phenomenon can be ascribed to a number of different reasons such as claustrophobic effects related to the physical characteristic of the environment; suggestions related to literature and movie memories and due to the fact that underground trains and stations offer ideal sceneries for suspense actions; automatisms often related to *train catch* experiences regardless a real condition of haste.

In movies, underground contexts often provide the scenery for thrilling actions, offering intricate paths with obstacles, corridors and stairs; the feeling of oppression related to the mouse trap suggestion and the metaphor of the labyrinth way-finding experience create the ideal set to communicate the prisoner escape run feeling. And, as documented by media news, in some cases underground stations can be dangerous, especially when and where the passenger frequency is lower.

As a synthesis, social interaction in any sense plays a crucial role in the final perception quality of the travel experience. In normal conditions of working, i.e. when the basic transport service functions are efficiently delivered, passengers' experience is significantly influenced by the presence of other travellers.

Crowded platforms and train vans are often associated to feelings of anxiety, nuisance, annoyance, distaste. In the presence of high people density, simple manoeuvres such boarding and getting out, taking place and getting a seat, offer the opportunity for competitive behaviours, little verbal aggressions, reproachful glance exchanging so causing irritation, frustration, or simply (and mostly) a careful attention in avoiding any kind of interaction with other people.

On the contrary, while the service fruition is obviously easier when few travellers are in the area, in less populated spaces the presence of other individuals is felt as something that must be checked up in order to avoid unpleasant events.

On the other hand, when people have the opportunity of positive social interaction in some form, from a simple exchange of smiles to a full conversation, including some speechless collaborative gestures, the flavour of the travel experience appears significantly improved (Fig. 7.9).

All this knowledge provided the basis for effective service design.

We want to point out that travellers' experience depends both on objective characteristics of the context, but also on the projection of mental frames associated to it. We can act according to the emotion produced by a physical environment, or modifying the mental attitude towards it. In the case of the underground stations, several advantages can be provided by the *adoption* of the place by the users.

7.3.3 Interpretation

So, we are able now to re-organise the observation results according to the model presented in Chap. 3.

Summarizing the results, we observed that:

(A) lone travellers tend to self-isolation, boredom, anxiety, with scarce attention to commercial ads or other visual events;
(B) lone travellers use survival strategies, creating: *readers, space invaders, hygiene fanatics, sleepers, chat catchers, workers, music listeners, phone users*;
(C) travellers joined to others, talk all the time, and showing a positive attitude and no anxiety at all;

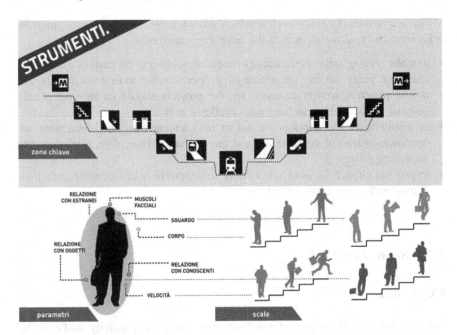

Fig. 7.9 An example of the final results produced during the observation phases of the project (The example is due to S. Acerbi, M. Arduini, P. Berardelli, M. Corradini). In the *upper part* of the image, the different locations of the underground system that were taken as a reference are represented; in the *lower* the scheme of the observed variables

(D) women change the position of their bags, and increase the walking speed;

(E) crowded environments increase anxiety and competition;

(F) positive interactions among travellers (a few words, smiles, and so on), also by chance, and between persons not knowing each other, reduce anxiety.

In terms of the model presented in the Chap. 3, we can say that;

- in general, *seeking* seems to be inhibited, or, at least, hidden; when present, it is watching for possible dangers (point D);
- boredom and anxiety, in particular in crowded environments, provide (latent) *rage* (points A and E);
- *fear* is largely present (points A, D and E);
- *panic/loss*, in terms of feeling themselves lonely, is present and hidden through any kind of activity (points A and B);
- *care* is limited to those groups of travellers talking each other (point C);
- *play* is a means used to reduce negative emotions, avoiding a confrontation with other travellers (point B).

It is to be noticed that weak positive signals (point F) reduce *rage*, *fear* and *panic/loss*, with emotionally positive effects.

The above remarks make evident which goals a project should pursue, in order to increase the positive aspects of the travelling experience:

- to make *seeking* arise; as we already observed, *seeking* is the basis of many other emotions, being the way for getting the proper attention to external signals, and it is a pleasant activity in itself; so, the projects should create perceptually unusual environment, catching the traveller's attention;
- to gently force the contacts (verbal or not) among travellers, increasing co-operation instead of competition, and then reducing *rage, fear, panic/loss* and increasing *care*;
- to provide chance for *play*, introducing some, possibly co-operative, activities, coping with some kinds of fun.

7.3.4 Some Proposals

7.3.4.1 Birds[4]

The declared goal of the project was to increase the *seeking* activity, and *care*, by influencing the behaviours of the travellers, leading them toward less noise, and recalling natural environments with wild life.

On a wall, the projection of a forest with birds recalls the natural environment, and some birds are singing on the trees; but, if the noise of the ambient grows, the birds fly away: as the higher the noise the less birds remain on the projection; they return back as the noise decreases (Fig. 7.10).

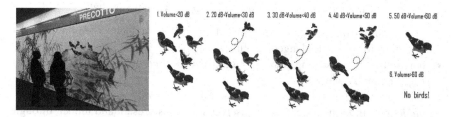

Fig. 7.10 Birds: a rendering of the proposal for a specific underground station in Milan, and some details on the technical interactions

---

[4] The project has been designed by Wei Dian and Zhu Chao.

Fig. 7.11 KindPoint: a storyboard explaining the concept, some ads warning the visitors of the existence of the initiative, and how it works, a prototyped scenario in a real underground station in Milan

7.3.4.2 KindPoint[5]

The project was designed apparently for helping foreign travellers, or in any case travellers inexperienced of the underground.

Let us put in the underground stations some bright orange spots on the floor, with proper signposts, explanatory posters and advertising campaign: if a person has some problems (where we are..., how to reach..., what to do for..., where is...), reaches one of those spots, standing inside; other persons understand he/she needs some helps, and are supposed to offer support.

The project, very simple, respecting all the given constraints, with very low installation costs and without execution costs, increases naturally the *seeking* activity, pushes people to interaction, increasing *care*, and then addressing most of the needs according to the above interpretation (Fig. 7.11).

---

[5] The project has been designed by S. Bonafini, E. Cattaneo and M. Ide.

Chapter 8
Future Developments

Abstract The chapter presents ongoing researches and studies related to future steps in designing emotions: many aspects of how to measure the raised emotions are discussed (functional Magnetic Resonance, Kansei engineering, automatic recognition of gestures, postures and face expressions, measures of physiologic parameters, etc.). Some new perspectives on designing emotions for interactive systems or services are discussed.

Keyword Emotion measure · Kansei engineering · Espression recognition · Posture recognition · Physiologic parameters for emotions · Emotional Interaction design

The field of Emotional Design is still young, and many further researches are on progress. We will here just make some remarks about the problem of the quantitative measures of emotions, and about the extension of the methods to Interaction Design.

8.1 Measuring Emotions

In the previous chapters we presented a model suggesting how perceptual aspects, simple like shapes, colours, rhythms and so on, and complex, based to semantic structures, analogies and metaphors, can influence the arousal of emotional status, and which emotions.

The goal is of course that of providing a designer of new information to project the emotions possibly felt by users: besides the evident impacts on the acceptance/non-acceptance in buying goods, there are other more important applications, in particular in health care and in healthy living.

If we are able to raise emotions in a health care environment, we can verify how much positively emotions can affect the effectiveness of the therapies, as many doctors suggest (Soresi 2005). The research path in this case should be the one shown in Fig. 8.1.

© The Author(s) 2015
M. Maiocchi, *The Neuroscientific Basis of Successful Design*,
PoliMI SpringerBriefs, DOI 10.1007/978-3-319-02801-9_8

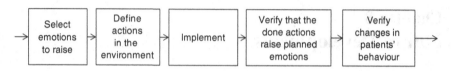

Fig. 8.1 A research path for emotional design in health care environments

Fig. 8.2 A fNRM of the
brain, with the evidence of the
excited parts

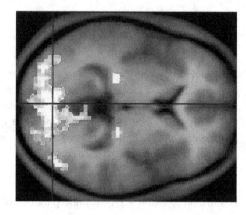

If the approach could be validated in terms of replicability and constancy of the
results, it could be applied for a re-definition of standard organisations in any health
care environment. Results have been locally obtained, as shown in the previous
chapter.[1]

But, how can we execute the fourth step, verification? The best way seems to be
the use of fNMR on the brain: according to the neuroscientific studies, emotion
arousal is related to the production of specific neurotransmitters, related to specific
brain areas; when one of those areas is activated, the local consumption of sugar
increases, i.e. the blood flow increases, and this can be tracked and properly
visualized (Fig. 8.2).

Unfortunately, beside the costs, the neuroimaging equipments are huge and
pervasive, and cannot be used extensively for our purposes.

There are alternative ways: among them, Kansei engineering methods, the
analysis of postures and face expressions, the use of local simple physiologic
parameters sensors.

---

[1] Of course we do not consider such approach as a substitutive method of medicine, but just as a
small contribution to the medical therapies effectiveness.

8.1.1 Kansei Engineering

Kansei is a Japanese word which means *senses, human preferences, feelings* etc. Kansei Engineering is a technology and methodology on design level or development level, which translates the "Kansei" of a person into a concrete product.

Kansei engineering data are generally coming from adjectives (kansei words) collected through interviews, to build the semantic space for design of products. The goal is the translation of human emotions into appropriate product design elements such as size, shape, colour, texture and so on.

Many experiences have been carried on, in particular for health care services. For example, studies have been carried on to map the selected adjectives on the Panksepp's emotional spaces; then, by interviewing the users, to verify which adjectives were considered applicable to some hospital waiting areas, comparing the results with the supposed proper profile of applicable adjectives; some examples are shown in the Figs. 8.3 and 8.4 (Shafieyoun 2014).

Seeking	Care	Play	Fear	Panic/Grief	Rage
Calm	Nice	Moving	Hateful	Depress	Agitate
Slow	Funny	Alive	Confusing	Boring	Heavy
Quiet	Beautiful	Active	Exiting	Sleepy	Chaotic
Apathetic	Friendly	Dynamic		Dozy	Lazy

Fig. 8.3 Mapping the Kansei words on the main Panksepp's emotions

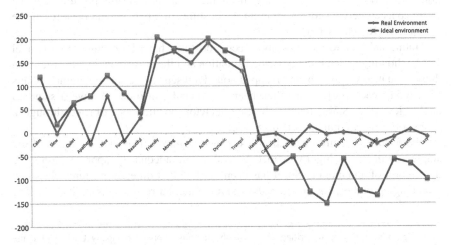

Fig. 8.4 The ideal profile (*squared dots*) of the adjectives representing a waiting area compared with the "measure" of a real case (*lozenges*)

8.1.2 Face Expressions, Postures

Human body is a very complex structure. Despite we pointed out just the brain as the site where emotions rise, the peripheral nervous system plays a very important role in the same function. For example, the sympathetic and the parasympathetic systems work as antagonists, allowing the fight/flight behaviours (Sapolsky 2004), and temporarily changing the physiological status (heart beat rate, blood pressure, breathing rhythm, immunitary capability, pain perception, thirst and hunger perception, fatigue endurance, and so on). In particular, many studies show the role of the *vagus* (the cranial nerve X, belonging to the parasympathetic system) (Porges 2011). The vagus nerve not only transmits signals to the peripheral parts of the body, but also collects information from them, reporting signals to the brain. In this function, it is responsible for many affective behaviours, exchanging signals with the limbic system, contributing to the production of oxytocin, and it is responsible of changes in the heart beat rate, in breath rhythm, in immobilization behaviours, in vocalization (and in particular for phatic expressions and for the prosody), in facial expressions and in some gestures and postures.

It is so evident that we could try to get information on some of the emotional status either by examining facial expressions, or by observing body postures.

The are many studies on the automatic recognition of facial expressions and the related emotions (Bettadapura 2012), and various commercial applications are available to recognize facial expressions and the related emotions; nevertheless, the concept of "emotions" in those studies and applications are quite far from the precise model provided by Panksepp.

Many researches are going on for an automatic recognition of the emotional status from the analysis of body postures (Kleinsmith 2007).

Some of them used very simple non intrusive technologies, to recognize emotions from body postures, starting from the idea that theatre postures can be assumed as archetypes for interpreting the postures of a person (and postures as the emotions[2]) (Radeta 2013).

Traditional theatre manuals about actor postures have been considered (Engel 1822), showing in detail the described situation and the corresponding posture; then those situations have been mapped on Panksepp's seven emotional status (Fig. 8.5a); each posture has then been "corrected", in order to have a standard angle view (Fig. 8.5b); finally, the raw skeleton position as extracted through a Microsoft Kinect equipment (Fig. 8.5c).

The postures schemes have been examined in terms of relative positions of the single segments, trying to extract general rules, adequately approximating the description of an emotion. So, it has been verified that different postures classified as belonging to the same emotion show similar angular relationships (Fig. 8.6). The

---

[2] The hypothesis has been verified also with the famous puppet's company Carlo Colla: the experts of the company confirmed that the posture is the most important element in representing emotions, in particular for puppets, unable to show facial expression.

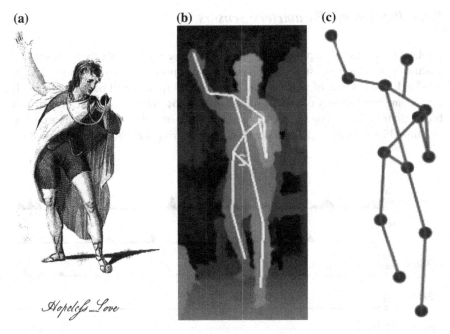

Fig. 8.5 **a** An example of a posture as described in traditional theatre manuals (*Hopeless Love*, we could map on Panic/Grief); **b** a schematic representation of the same posture, seen in front to the subject; **c** the corresponding scheme extracted through Kinect

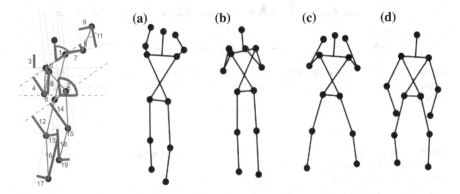

Fig. 8.6 The angles taken under consideration in the model, and four postures for *Rage*

validation is still on progress, but some preliminary results are encouraging, allowing the idea that a non-intrusive observation can record the emotional status of few persons simultaneously (think for example, of the use in measuring the emotional status in a waiting area).

8.1.3 Physiologic Parameters Sensors

Further researches are on progress (Radeta 2014) to verify the usability of sensors for evaluating the emotional status, as suggested by the *polyvagal theory* (Porges 2011).

The declared status of persons watching selected movies will be compared with the dynamic behaviours of skin humidity, blood flow and pressure, heart beat and breath rhythm. The set up samples should allow to verify the instantaneous arousing emotion (Fig. 8.7), as well as the prevailing ones during the whole movie (Fig. 8.8).

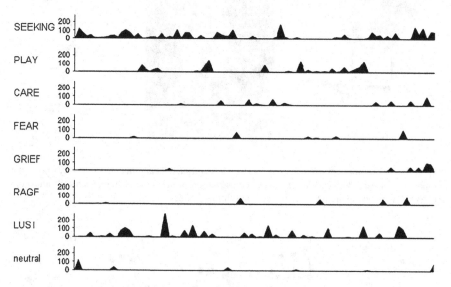

Fig. 8.7 The local distribution of the emotion feeling as evaluated by a group of watchers while attending to the movie *Nine 1/2 Weeks* by Adrian Lyne. With Mickey Rourke, Kim Basinger (1986)

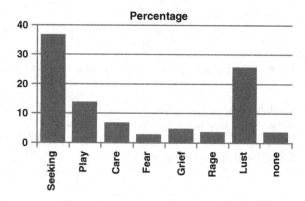

Fig. 8.8 The average distribution of the emotion feeling as evaluated by the same group of watchers while attending to *Nine 1/2 Weeks*

8.2 Emotional Interaction Design

Interaction with artefact can be classified according their dynamic properties, and this opens a new chapter we can call Emotional Interaction Design.

Interaction can be:

1. *Static* (pattern): a hammer is and will be just a hammer; maybe that I will interpret in different ways according to the context (i.e. the structures I use for its interpretation, but it is matter of interpretation, not a property of the artefact); the same for colours, shapes, physical properties;
2. *Fixedly dynamic*: the sequences of commands to be provided to a sequential machine (drink vendor, cash dispenser, a software user interface, etc.) present a script never changing in time; the operations are characterised by synchronicity (in some case there is some parallelism in actions, with synchronisation points); for example, the dialogue a user has with a new TV during the automatic tuning is absolutely determined: the user starts with a command, the command can provide some choices to be selected by the user, then it can start with actions performed in a fixed sequence, possibly with the issue of many messages, and during the execution of those tasks the user can do other jobs in parallel, until a new synchronisation point is reached, in which the user must interact in the only predefined way; the same concerns the installation of new software programs, and many others; from a mathematical point of view, those artefacts (and the related scripts) can be modelled with combinatorial machines, with finite state automata, with automata with memory); this model apply to any automatic service;
3. *Openly dynamic*: the sequences of messages exchanged between the user and the "artefact" cannot be pre-defined, and many unpredictable situations can occur; this is typically what happens with a service human operated, such as a call centre service, or a bank window, or a shop, in which the answering procedures, even if deeply examined and designed, cannot consider *a priori* all the possible occurrences[3]; in this case, the interaction is typically diachronic; mathematical models require the representation of concurrent processes, typically through Petri Nets.

Those three models correspond to different Interaction Design levels:

1. the first is related to very few interaction problems, usually reduced to ergonomics, usability and affordance (besides the emotional aspects we could add);
2. the second case can be studied from the point of view of the communication channels (e.g. keyboard, touch, vocal, etc.) and of the metaphors in the commands (e.g. drag and drop, select, copy and paste, folder, etc.); emotions can be

---

[3] In fact, there is a hard design problem: it is quite easy to define the behaviours of a "machine" in front of any allowed behaviour of the user, but is impossible define all the unexpected (and then forbidden) behaviours of a user.

affected by characteristics of the channel and of the "static" element composing the dialogue (e.g. the capability of selecting a voice gender for a car satellite navigator is a very poor trial to introduce emotional impacts, as well as the use of courtesy forms in the dialogue on a screen, or similar);

3. the third case, typically a service run by humans, is richer, more complex and full of uncertainty;

- *rich*, because a metaphoric script can be over imposed, adding a specific style to the dialogue;
- *complex*, because the communication is twofold: when a user gives a message to the server, the latter get not only the information, but also the emotional status of the user; his/her answer will contain both kind of messages, and so on; moreover, the emotional status of the other party might be one of the possible design goals (the customer should be happy, while the customer could sometimes try to overwhelm the server, a third communication channel is naturally built, the user feeling his/her status, and the same for the server) (Maiocchi et al. 2013);
- *uncertain*, because it is not possible to list *a priori* all the possible cases.

There are very few researches oriented to this kind of development. One of them, carried on as a master thesis (Alimohammadi 2013) used the following approach:

- interpret the interaction with a service (human or machine driven) as an interaction between two persons playing different roles;
- examine from theatre, literature and movies sample dialogues, in which the different roles are evident, as well as the emotions characterising the dialogue;
- classify the dialogues according to Panksepp's seven emotions;
- extract the perceptual elements of the dialogue, able to raise such emotions (prosody, language, rhetoric, gestures, postures, expressions, and so on);
- examine real cases of services and the related reactions of the users[4];
- try a set of suggestion for a new "interaction rhetoric".

The study examined real situations, such as the behaviour and the reaction of elder users of automatic payment systems in supermarkets or in train stations. The results of the work should be considered as preliminary, but could possibly show a way to start deeper studies.

Further researches examined the way in which emotions work within the interaction while playing videogames, and in particular for multiplayer videogames. In fact, when two or more persons are playing together, the emotions are not simply

---

[4] In this phase significant results emerged: most of the disliked interaction was asymmetric (i.e. the roles of the user and of the provider were at the same level); moreover, the user was not the "chief" of the process, but was just suffering the situation; moreover, the played roles were comparable to the schemes severe father–son , chief–soldier, chief–low level worker; who can give orders—who can only execute.

raised by events or behaviours, but also (possibly mainly) by mirroring the emotions of other players (Maiocchi 2012). From the research emerged the opportunity to model the process of emotions in playing through co-routines, going on in parallel with the concurrent player's processes.

References

M. Alimohammadi, *Verso Emotional Interaction Design*, Master thesis, Politecnico di Milano, Master in Communication Design, 2013

B. Archer, *The Need for Design Education* (Royal College of Art, London, 1973)

M.M. Aslam, Are you selling the right colour? a cross-cultural review of colour as a marketing cue. J. Mark. Commun. **12**, 15–30 (2006)

G. Bateson, *Steps to an Ecology of Mind: Collected Essays in Anthropology, Psychiatry, Evolution, and Epistemology* (University of Chicago Press, Chicago, 2000) (reprint. First published 1972)

V. Bettadapura, *Face Expression Recognition and Analysis: The State of the Art*, Technical Report, 1–27, Georgia Institute of Technology, 2012

H. Beyer, K. Holtzblatt, *Contextual Design: Defining Customer-Centered Systems* (Morgan Kaufmann, San Francisco, 1998)

A. Branzi, *Gli strumenti del design non esistono. La dimensione antropologica del design*, Johan and Levi, 2013

B.E. Bürdek, *Design: The History, Theory and Practice of Product Design* (Birkhaeuser Basel, Boston, 2005)

L.L. Cavalli Sforza, *Cultural Transmission and Evolution: A Quantitative Approach* (Princeton University Press, Princeton, 1981)

L.L. Cavalli Sforza, *L'evoluzione Della Cultura*, Codice, 2008

J. Cooper, *Cognitive Dissonance: 50 Years of a Classic Theory* (SAGE, London, 2007)

G. Cox, Cox *Review of Creativity in Business: Building on the UK's strengths*, http://www.hm-treasury.gov.uk/cox—nov 2005

C. Darwin, *The Expression of the Emotions in Man and Animals* (John Murray, London, 1872)

J. Diamond, *Guns, Germs, and Steel: The Fates of Human Societies* (W. W Norton, New York, 1999)

P. Ekman, *Unmasking the Face: A Guide to Recognizing Emotions from Facial Expressions* (Malor Books, Cambridge, 2003)

J.J. Engel, H. Siddons, *Practical Illustrations of Rhetorical Gesture and Action* (Sherwood, Neely and Jones, London, 1822)

K. Erwin, *Communicating the New* (Wiley, New York, 2014)

ET, Treccani.it, L'enciclopedia Italiana, http://www.treccani.it/enciclopedia/disegno-industriale/ (As in November 2014)

L. Festinger, *A Theory of Cognitive Dissonance* (Stanford University Press, Stanford, 1957)

A. Frutiger, *Symbols Signs* (Niggli Verlag, Sulgen CH, 2009)

M.D. Gershom, *The Second Brain* (HarperCollins Publishers, New York, 1988)

G. Hofstede, G.J. Hofstede, M. Minkov, *Cultures and Organizations: Software of the Mind*, 3rd edn. (McGraw-Hill, USA, 2010)

ICSID International Council of Societies of Industrial Designers (2004). Accessed 27 Aug 2004

W. Kandinsky, *Point and Line to Plane,* 1st edn. (Dover Publications, New York, 1947)

© The Author(s) 2015
M. Maiocchi, *The Neuroscientific Basis of Successful Design*,
PoliMI SpringerBriefs, DOI 10.1007/978-3-319-02801-9

G. Kanizsa, *Grammatica del vedere. Saggi su percezione e Gestalt,* Il Mulino, 1997

A. Kleinsmith, N. Berthouze, *Recognizing Affective Dimensions from Body Posture.* Lecture Notes in Computer Science, vol. 4738 (Springer, Berlin, 2007), pp. 48–58

W. Köhler, *Gestalt Psychology* (Liveright, New York, 1929)

G. Lakoff, M. Johnson, *Metaphors We Live* (University of Chicago Press, Chicago, 1980)

G. Lakoff, *Don't Think an Elephant* (Chelsea Green Publishing, Chelsea, 2004)

M. Lindstrom, *Buyology: Truth and Lies About Why We Buy* (Broadway Books, New York, 2010)

X. Lu, P. Suryanarayan, R.B. Jr Adams, J. Li, M.G. Newman, J.Z. Wang, On shape and the computability of emotions, in MM'12, Nara, Japan, Oct 29–Nov 2, 2012 (ACM 978-1-4503-1089)

M. Maiocchi (ed.), *La comunicazione emozionale negli ambienti ospedalieri,* Maggioli, 2007

M. Maiocchi (ed.), *Design e Comunicazione per la Sanità,* Maggioli, 2008

M. Maiocchi, M. Pillan, *Design e Comunicazione,* Alinea, 2009

M. Maiocchi (ed.), *Design e Medicina,* Maggioli, 2010

M. Maiocchi (ed.), *Artisti e Salute,* Istituto dei Tumori di Milano, 2010a

M. Maiocchi, M. Pillan, I. Suteu, *Monkey Business: the Academic World* (ICERI, Madrid, 2011)

M. Maiocchi, M. Pillan, M. Radeta, P. Righi Riva, Mirror emotions and sex games, in *Arse Elektronica Conference,* San Francisco, 2012

L. Margulis, D. Sagan, *Acquiring Genomes: A Theory of the Origins of Species* (Basic Books, New York, 2003)

E. Mari, *Progetto e Passione* (Bollati Boringhieri, Turin, 2001)

B. Munari, *Fantasia, Invenzione,creatività e immaginazione nelle comunicazioni visive*, Laterza, 1977

B. Munari, *Da cosa nasce cosa*, Laterza, 1981

B. Munari, *Le macchine di Munari*, Corraini, 2001

D. Norman, *Design emozionale*, Apogeo, 2004

Y. Oh, Métodos para a criatividade emocional, in Cadernos de Estudos Avançados em Design—Emoção, vol. 8—Ed. UEMG—Universidade do Estado de Minas Gerais 2013

V. Packard, *The Hidden Persuaders* (David McKay Co., Inc, New York, 1957)

J. Panksepp, L. Biven, *The Archeology of the Mind* (W. W. Norton and Company, New York, 2012)

V. Papanek, *Design for the Real World: Human Ecology and Social Change* (Pantheon Books, New York, 1971)

R. Pierantoni, La trottola di Prometeo. Introduzione alla percezione acustica e visiva, Laterza, 1996

R. Plutchik, *Emotions and Life: Perspectives from Psychology, Biology, and Evolution* (American Psychological Association, Washington, 2002)

S. Porges, *The Polyvagal Theory: Neurophysiological Foundations of Emotions, Attachment, Communication, and Self-Regulation* (W. W. Norton, New York, 2011)

V.J. Propp, *Morphology of the Folktale* (University of Texas Press, Austin, 1928)

M. Radeta, M. Maiocchi, Towards automatic and unobtrusive recognition of primary-process emotions in body postures, in *Proceedings of Conference on Affective Computing and Intelligent Interaction (ACII)*, Geneve, Switzerland, 2013

M. Radeta, Z. Shafieyoun, M. Maiocchi, Affective timelines. towards the primary-process emotions of the movie watchers measurements based on a self-annotation and affective neuroscience, in *International Conference on Design and Emotion*, Universidad de los Andes, Bogotá, 2014

V. Ramachandran, W. Hirstein, The science of art. Conscious. J. **6**, 15–51 (1999)

V.S. Ramachandran, E.M. Hubbard, Synaesthesia: a window into perception, thought and language. J. Conscious. Stud. **8**(12), 3–34 (2001)

G. Rizzolatti, C. Sinigaglia, *Mirrors In The Brain: How Our Minds Share Actions and Emotions* (Oxford University Press, New York, 2008)

L.D. Rosenbaum, *See What I am Saying. The Extraordinary Power of Our Five Senses* (W. W. Norton, New York, 2010)

R. Sapolsky, *A Primate's Memoir* (Touchstone Books, New York, 2002)

R. Sapolsky, *Why Zebras Don't Get Ulcers* (Owl Books, New York, 2004)

M. Sclavi, Arte di ascoltare e mondi possibili. Come si esce dalle cornici di cui siamo parte, Bruno Mondadori, 2003

Z. Shafieyoun, M. Radeta, M. Maiocchi, *Environmental Effect on Emotion in Waiting Areas Based on Kansei Engineering, KEER* (Linköping University, Sweden, 2014)

P. Sica, H., Bioledge. A model for everything, a method for design, Master thesis in Communication Design, Politecnico di Milano, 2011

E. Soresi, *Il Cervello Anarchico*, UTET, 2005

G. Vallortigara, *Cervello di gallina*, Bollati Boringhieri, 2005

Van T. Gorp, E. Adams, *Design for Emotions* (Morgan Kaufman, Boston, 2012)

T.W. Whitfield, T.J. Whiltshire, Color psychology: a critical review. Genetic, Social, General Psychology Monographs, **116**. 387–412 (1990)

S. Zeki, *Splendour and Miseries of the Brain. Love, Creativity and the Quest for Human Happiness* (Wiley, Chichester, 2009)

Name Index

© The Author(s) 2015
M. Maiocchi, *The Neuroscientific Basis of Successful Design*,
PoliMI SpringerBriefs, DOI 10.1007/978-3-319-02801-9

Subject Index

© The Author(s) 2015
M. Maiocchi, *The Neuroscientific Basis of Successful Design*,
PoliMI SpringerBriefs, DOI 10.1007/978-3-319-02801-9

Printed in the United States
By Bookmasters